Studies in Autonomic, Data-driven and Industrial Computing

Series Editors

Swagatam Das, Indian Statistical Institute, Kolkata, West Bengal, India

Jagdish Chand Bansal, South Asian University, Chanakyapuri, India

The book series Studies in Autonomic, Data-driven and Industrial Computing (SADIC) aims at bringing together valuable and novel scientific contributions that address new theories and their real world applications related to autonomic, data-driven, and industrial computing. The area of research covered in the series includes theory and applications of parallel computing, cyber trust and security, grid computing, optical computing, distributed sensor networks, bioinformatics, fuzzy computing and uncertainty quantification, neurocomputing and deep learning, smart grids, data-driven power engineering, smart home informatics, machine learning, mobile computing, internet of things, privacy preserving computation, big data analytics, cloud computing, blockchain and edge computing, data-driven green computing, symbolic computing, swarm intelligence and evolutionary computing, intelligent systems for industry 4.0, as well as other pertinent methods for autonomic, data-driven, and industrial computing.

The series will publish monographs, edited volumes, textbooks and proceedings of important conferences, symposia and meetings in the field of autonomic, data-driven and industrial computing.

Parikshit Narendra Mahalle · Nancy Ambritta P. ·
Sachin R. Sakhare · Atul P. Kulkarni

Foundations of Mathematical Modelling for Engineering Problem Solving

Springer

Parikshit Narendra Mahalle
Department of Artificial Intelligence and Data Science
Vishwakarma Institute of Information Technology
Pune, India

Sachin R. Sakhare
Department of Computer Engineering
Vishwakarma Institute of Information Technology
Pune, India

Nancy Ambritta P.
Glareal Software Solutions PTE. Ltd.
Singapore, Singapore

Atul P. Kulkarni
Department of Mechanical Engineering
Vishwakarma Institute of Information Technology
Pune, India

ISSN 2730-6437 ISSN 2730-6445 (electronic)
Studies in Autonomic, Data-driven and Industrial Computing
ISBN 978-981-19-8827-1 ISBN 978-981-19-8828-8 (eBook)
https://doi.org/10.1007/978-981-19-8828-8

This Springer imprint is published by the registered company Springer Nature Singapore Pte Ltd.
The registered company address is: 152 Beach Road, #21-01/04 Gateway East, Singapore 189721, Singapore

Preface

Scientists investigate that which already is ... Engineers create that which has never been.

—Albert Einstein

Due to the transformation to Industry 4.0 and emergence of Society 5.0, there is a need of reliable problem solving and mathematical modeling. This will enable the support to advance architectures and technologies resulting into the optimal and efficient solutions. This book aims at introducing fundamentals of problem solving, need of mathematical modeling and the importance of mathematical modeling for different categories of the problems to undergraduate and postgraduate students and researchers in science and engineering. The text is divided into three parts: concepts, mathematical theories and case studies in different domains of engineering problem solving.

Research and development plays a major role in the economic growth and the GDP of a nation. Hence, every nation takes steps toward the improvement of the Gross Enrollment Ration (GER) under the advisory of the National Knowledge Commission, thereby increasing the funding in support of admissions to postgraduation (PG) and Ph.D. programs. In these PG and Ph.D. programs, mathematical modeling becomes an integral part of research. However, there is a scarcity of good literature for engineering problem solving. This book aims at providing an insight into mathematical modeling for engineering problem solving irrespective of the branch of study, thereby making it worthwhile for the researchers.

This book fundamentally aims at improving the mathematical modeling skills among the users by enhancing the ability to understand, connect, apply and use the mathematical concepts to the problem at hand. This book provides the readers with an in-depth knowledge of the various categories/classes of research problems that professionals, researchers and students might encounter following which the application of appropriate mathematical models is explained with the help of case studies. This book is majorly concerned in developing the expertise of identifying the category or class into which the problem at hand falls, which eventually makes the modelling aspect easier. This book is targeted at academicians, researchers, students and professionals who belong to all engineering disciplines.

This book is aimed at providing timely and useful content to the mass readers, especially to researchers and Ph.D. scholars to whom mathematical modeling becomes inevitable, irrespective of the branch of study.

The main characteristics of this book are:

- Provides a strong foundation for the readers regarding the basics of mathematical modeling, its relevance and need.
- Provides illustrations and pseudo-code for better understanding and applicability.
- Elaborates on the problem solving and mathematical modeling process.
- Equips the readers with the ability to identify and categorize the problem at hand (specifically regarding computer engineering), thereby making it an effortless task for the users to arrive at an appropriate mathematical model for representing the problem.
- Provides an in-depth understanding on the various categories/classes of problems (dedicated chapter for each category of problem).
- Provides numerous case studies for providing a clear understanding of concepts in real time and serves as a guide to aspiring students/professionals by presenting various research openings in the field.

This book is specifically designed for beginners who want to get acquainted with the concept of mathematical modelling and the various categories. Focus on identification of the category of problem and the application of appropriate model is a key highlight in this book. The book is useful for undergraduates, postgraduates, industry, researchers, teachers and research scholars in engineering, and we are sure that this book will be well received by all stakeholders.

Pune, India — Parikshit Narendra Mahalle
Singapore — Nancy Ambritta P.
Pune, India — Sachin R. Sakhare
Pune, India — Atul P. Kulkarni

Contents

1 Introduction ... 1
1.1 Modeling ... 1
1.2 Mathematical Modeling ... 2
1.3 General Steps ... 7
1.4 Trends in Teaching and Learning ... 9
1.5 Summary ... 9
References ... 10

2 Problem Solving and Mathematical Modeling ... 11
2.1 Problem Evolution ... 11
2.2 Problem Solving ... 13
2.3 Problem Classification ... 17
2.4 Modeling Process and Teaching Approaches ... 19
2.5 Summary ... 22
References ... 22

3 Decision Problems ... 23
3.1 Introduction ... 23
3.1.1 Control Theory ... 23
3.1.2 Game Theory ... 25
3.1.3 Probability and Statistics ... 28
3.1.4 Multicriteria Decision Analysis ... 32
3.2 Motivation ... 33
3.3 Case Studies ... 34
3.4 Summary ... 37
References ... 38

4 Optimization Problems ... 39
4.1 Introduction ... 39
4.1.1 Mathematical Modeling ... 39
4.1.2 History Development of Optimization ... 40
4.1.3 Optimization ... 41

4.2 Motivation ... 43
4.3 Case Study ... 43
4.4 Summary ... 55
References ... 55

5 Delay Problems ... 57
5.1 Introduction ... 57
5.1.1 Queuing Theory ... 58
5.1.2 Delay Differential Equations ... 66
5.2 Motivation ... 82
5.3 Case Studies ... 82
5.4 Summary ... 86
References ... 86

6 Data Science Problems ... 87
6.1 Introduction ... 87
6.1.1 Data Analytics and Learning Methodologies ... 101
6.1.2 Data Visualization Tools and Data Modeling ... 104
6.2 Motivation ... 136
6.3 Case Studies ... 138
6.4 Summary ... 140
References ... 140

7 Pandemic Problems ... 143
7.1 Introduction ... 143
7.1.1 Impacts of Pandemics ... 145
7.2 Motivation ... 145
7.2.1 Types of Epidemic Models ... 147
7.3 Case Studies ... 150
7.3.1 Compartmental Models ... 150
7.3.2 COVID-19 Pandemic Problem ... 151
7.4 Summary ... 153
References ... 154

8 Interdisciplinary Engineering Problems ... 157
8.1 Introduction ... 157
8.2 Motivation ... 158
8.3 Case Studies ... 159
8.4 Summary ... 165
References ... 166

9 Conclusion ... 167
9.1 Summary ... 167
9.2 Research Openings and Future Outlook ... 168

About the Authors

Dr. Parikshit Narendra Mahalle obtained his B.E. degree in Computer Science and Engineering from Sant Gadge Baba Amravati University, Amravati, India, and M.E. degree in Computer Engineering from Savitribai Phule Pune University, Pune, India. He completed his Ph.D. in Computer Science and Engineering specialization in Wireless Communication from Aalborg University, Aalborg, Denmark. He was Postdoc Researcher at CMI, Aalborg University, Copenhagen, Denmark. He worked as Professor and Head in the Department of Computer Engineering at STES's Smt. Kashibai Navale College of Engineering, Pune, India, since 2005 to June 2021. Currently, he is working as Professor and Head in Department of Artificial intelligence and Data Science, Vishwakarma Institute of Information Technology, Pune, India. He has more than 21 years of teaching and research experience. He is serving as Subject Expert in Computer Engineering, Research and Recognition Committee at several universities like SPPU (Pune), SGBU (Amravati). He is Senior Member IEEE, ACM Member, Life Member CSI and Life Member ISTE. Also, he is Member of *IEEE transaction on Information Forensics and Security*, *IEEE Internet of Things Journal*. He is Reviewer for IGI Global—committee member for international conferences and symposium like IEEE ICC, IEEE INDICON, IEEE GCWSN, IEEE ICCUBEA, etc. He is Reviewer for the Springer *Journal of Wireless Personal Communications*, Reviewer for Elsevier *Journal of Applied Computing and Informatics*, Member of the Editorial Review Board of IGI Global—*International Journal of Ambient Computing and Intelligence (IJACI)* and Member of the Editorial Review Board for *Journal of Global Research in Computer Science*. He has published more than 190 research publications having 1810 citations and H-index 18. He has seven edited books to his credit by Springer and CRC Press. He has seven patents to his credit. He has also delivered invited talk on "Identity Management in IoT" to Symantec Research Lab, Mountain View, California. He has delivered more than 100 lectures at national and international level on IoT, big data and digitization. He has authored 15 books on subjects like *Context-aware Pervasive Systems and Application* (Springer Nature Press), Design and Analysis of Algorithms, *Cambridge International Journal of Rough Sets and Data Analysis (IJRSDA)*, Associate Editor for IGI Global—*International Journal of Synthetic Emotions (IJSE)*, *International*

Journal of Grid and Utility Computing (IJGUC). He is Member—Editorial Review Board for IGI Global—*International Journal of Ambient Computing and Intelligence (IJACI)*. He is also working as Associate Editor for IGI Global—*International Journal of Synthetic Emotions (IJSE)*. He has also remained in a technical programme university, *Identity Management for the Internet of Things* (River Publications), *Data Structure and Algorithms* (Cengage Publications) and *Programming using Python* (Tech—Neo Publications MSBTE). He had worked as Chairman of Board of Studies (Information Technology), SPPU, Pune. He is working as Member—Board of Studies (Computer Engineering), SPPU, Pune. He has been Member of the Board of Studies at several institutions like VIT (Pune), Government College (Karad), Sandeep University (Nashik), Vishwakarma University (Pune), Dr. D. Y. Patil International University (Pune), etc. He has also remained Technical Programme Committee Member for many international conferences. He is a recognized Ph.D. guide of SSPU, Pune, and guiding seven Ph.D. students in the area of IoT and machine learning. Recently, two students have successfully defended their Ph.D. He is also Recipient of "Best Faculty Award" by Sinhgad Institutes and Cognizant Technologies Solutions. His recent research interests include algorithms, Internet of things, identity management and security. He has visited a few countries like Denmark, France, Sweden, Germany, Austria, Norway, China, Switzerland and Singapore.

Nancy Ambritta P. graduated in Computer Science and Engineering from Anna University, Tamil Nadu, India, in the year 2010 and has her master's in computer engineering from Smt. Kashibai Navale College of Engineering, Pune, India, in the year 2015. Her research interests are cloud security and the future Internet. She has about 3 years of experience in the software engineering field and 3.5 years of experience in teaching and research. She has published research articles in international journals and conferences to her credit. She has also authored a book titled *The Convergence of Internet of Things and Cloud for Smart Computing*. She is Reviewer for the Springer Journal of *Wireless Personal Communications*.

Dr. Sachin R. Sakhare has obtained his doctorate in Computer Science and Engineering in 2014. He has 25 Years of experience, and currently, he is working as Professor in Computer Engineering at Vishwakarma Institute of Information Technology, Pune, India.

He is Member of Computer Society of India and ISTE. He has availed the research workshop grants from BCUD, Pune University. He has published 32 research communications in Web of Science and Scopus indexed journals and presented papers in national and international conferences, He has 202 citations and H-index four. He has authored four books. He has filed and published four patents and three copyrights. He is invited as Member of Technical and Advisory Committee for various international conferences.

He is and currently guiding four Ph.D. scholars at Savitribai Phule Pune University. He worked as Reviewer of *Applied Soft Computing journal* from Elsevier, *The*

Journal of Soft Computing from Springer, IEEE-ICCUBEA-2018 and 2019, IEEE-INDICON, Springer and ACM conferences. Dr. Sachin has delivered invited talks at various conferences, FDPs and STTPs as well as to PG students.

He is associated with international universities also. He was invited as Guest Faculty at Amsterdam University of Applied Sciences (AUAS), the Netherlands, to conduct sessions on data structures and Internet of things. He has also visited Aalborg University, Copenhagen, Denmark, and RWTH Aachen University, Germany, for academic collaboration.

Dr. Atul P. Kulkarni works as Associate Professor in the Department of Mechanical Engineering, at Vishwakarma Institute of Information Technology, Pune. He as about 6 years of experience in the industry and about 20 years of experience in the teaching field. He graduated in Bachelor of Engineering from the North Maharashtra University in 1995 and M.Tech. from Dr. Babasaheb Ambedkar Technological University, Lonere, in 2006. He has also been awarded his Ph.D. from Dr. Babasaheb Ambedkar Technological University, Lonere, in 2016. He has published one book and about 65 research articles (journals and conference papers) in reputed journals. He has five patents and six copyrights to his credit. He has guided many UG and PG students and led to successful completion of their projects.

His current research interests include PVD coatings, FEA for machining application, modelling and machining of exotic materials, composite materials, composite materials and analysis of manufacturing processes of historical monuments.

Chapter 1
Introduction

1.1 Modeling

Data, understanding, information, knowledge and wisdom are main prerequisites for the structured design and effective implementation of any system. It is very difficult to troubleshoot any tool or system unless one does not understand how fundamentally it works and every functional block of the system is built. Problem solving on paper and by computers both need fundamental understanding of underline engineering problem irrespective of the branch or domain it belongs. Essentially, the problem solving by computers requires the entire system to be represented first as black box with the abstract representation of system and then the white box with detailed representation of the system [1]. Empirical means is one of the ways to gain this understanding through observations and series of experiments. However, empirical means and information derived empirically are not always sufficient and it is also not suitable for all the use cases in engineering problem solving. Theoretical analysis of the given problem is yet another alternative approach which includes the formal analysis of the entire system for solving. However, this approach requires building the comprehensive framework with required assumptions and the framework need to be verified before one gets into the system design.

However, in empirical and theoretical analysis, making assumptions for system parameter is very crucial. It is also possible that the system aspects in this analysis while observations ad experiments repeats over periodic or continuous intervals. In some cases, this occurrence of system aspects is not witnessed at all. These irregularities and uncertainties cannot be expressed easily in such analysis, and there should be a provision of model building. In the sequel, modeling is the best way before one gets into the solution domain. Modeling is an abstract representation of any entity under discussion. It essentially converts all required aspects of the entity or problem which are required to represent the overall process. However, model generally does not include the implementation or algorithmic details. The main aim of abstraction is to make the entire system presentable with respect to the input, output and all functional blocks. The model of township deployed at the entrance of residential

P. N. Mahalle et al., *Foundations of Mathematical Modelling for Engineering Problem Solving*, Studies in Autonomic, Data-driven and Industrial Computing,
https://doi.org/10.1007/978-981-19-8828-8_1

project is the appropriate example of model which depicts how the entire project will look like including the list of all amenities, organization of the project, internal roads, layout, etc., as a bigger picture. Model always helps to represent the essential aspects or features of the system, behavior, as well as the different states.

In the real world, building abstract representation in terms of model is always affordable in terms of cost and also safer for real-world use cases. Due to easy portability (crisp and concise nature) of model, it is very convenient to test, validate and analyze the model before one goes for development and deployment. Testing is very important phase in any application development life cycle and modeling enables better functional testing to detect unanticipated issues and they can be fixed before the development. This alleviates the human efforts required in change management.

1.2 Mathematical Modeling

There are various definitions and perspectives of mathematical modeling in the literature [2–4]. Every definition is coined with different perspective and considers different problem in engineering. When the mathematics or mathematical tools are used for abstract representation of any entity or process, then the model is referred as mathematical model. Mathematical formulation or mathematical equation is used in mathematical modeling to depict required features of entire system or process. Mathematical model also establishes the mathematical relationship between inputs and outputs and elaborates more on the entire process. It is very important to note that the mathematical modeling can be done for the given problem as well as for solution to the given problem. In mathematical equation, the dependent variable represents the features or characteristics used for describing the state and behavior of the entire process. In contrast, the independent variables are those which can be changed continuously or periodically in order to analyze its effect on other dependent variables. As a bigger picture in general, mathematical model in terms of dependent and independent variable can be represented as:

$$\mathrm{DV} = f(\mathrm{IV},\ \mathrm{IP},\ \mathrm{AF}) \tag{1.1}$$

where

DV = Dependent variable
IF = Independent variable
IP = Input parameters
AF = Acting forces

This representation of mathematical modeling clearly indicates that it will take different forms based on the underlined engineering problem. The concept of variable plays crucial role in building the mathematical models. The set of dependent and independent variable and their nature changes based on the type of engineering problem, and it is elaborated in Chapter 2 of this book. There are lot of misunderstanding and fear in the readers toward building the mathematical models of the problems essentially in the engineering domain. Depicting quantitative relation or calculations in the real world using mathematical tool, expression or language is the first step in mathematical modeling. In this step, variables play a crucial role to build the generic model. Following examples throw more light on it.

Example 1

Problem: Atul works in the company at very senior position and his pay bracket comes in the tax rate of 40%. What is the amount of tax Atul needs to pay?

Solution:

Input required to build the generic mathematical model is the annual salary of Atul. If we do not know this parameter, then we can replace it with a variable. Let's define the variable S for salary.

S = Atul's annual salary.

Atul will have to pay 40% of this annual salary in taxes, hence.

Payable Tax = 40% S = 0.40 S.

Example 2

Problem: Sachin is a very talented stock trader and always interested to invest in a stock. His target is to have returns of 15% annually. How much money will Sachin have at the end of the financial year?

Solution:

In this case, we do not know Sachin's initial investment. We can define it with variable I.

I: the amount invested by Sachin in the stock market.

Hence, the amount returned to Sachin at the end of financial year will be.

Returns = I + (10%) I.

Example 3

Problem: Generate set of natural numbers giving remainder one when divided by three.

Solution:

Let S_1 be the set of natural numbers which represent infinite set and denoted as:

$$S_1 = \{1, 2, 3, 4, 5, 6, \ldots\}$$

Let S_2 be the set of natural numbers giving remainder one when divided by three.

$$S_2 = \{1, 4, 7, 9, 13, \ldots\}$$

It can be seen that $S_2 \subseteq S_1$. In line with this, the initial condition is as follows:

$$A_0 = 1$$

$$A_{n+1} = A_n + 3$$

The final solution is:

$$\text{An} = \begin{cases} 1 \text{ if } n = 0 \\ \text{An} + 1 - \text{An} + 3 \end{cases}$$

Example 4

Consider the construction project in civil engineering which has three independent phases. The meaning of independent phases means one phase cannot be initiated before the previous phase is completed. Every phase has the cost associated with it which includes fixed cost and variable cost. The fixed cost is independent of the duration of that phase and the variable cost is based on the duration of each phase. The contractor is supposed to quote the price for entire project ensuring the profit of 10% on the total cost. The objective is to build the mathematical model to present total cost of the project and the price for each phase with respect to the duration of different phases. Table 1.1 provides the initial assessment of the project.

Phases	1	2	3
Cost in Rs (Fixed Component)	2,000,000	3,000,000	4,000,000
Cost in Rs/Day (Variable Component)	15,000	14,000	16,000

Solution:

The duration of each phase is important component on which the variable cost depends. Let us define three different variables for different phases to indicate the duration in days.

Table 1.1 Parameter matrix for mathematical modeling

Sr.No.	Domain	Entity	Functional component	Mathematical parameter
1	Chemical Engineering	Reactors	Chemical bonds and conservation of mass	Mass balance and inputs
2	Civil Engineering	Structure	Momentum management	Force
3	Mechanical Engineering	Machine	Torque laws	Forces and equilibrium
4	Electrical Engineering	Circuit	Conservation of charge	Current balance
5	Computer Engineering/Information Technology/Artificial Intelligence	Data	Association, dependency	Inputs

P_1 = Phase 1 duration
P_2 = Phase 2 duration
P_3 = Phase 3 duration

As mentioned in the given problem, the cost of every phase is divided into two components, i.e., fixed cost and the variable cost per day, and every phase has different duration in terms of number of days. In the sequel.

If the phase 1 duration is P_1 days, the resultant cost for this phase is given by:

$$\text{Cost}_{\text{Phase 1}} = 2,000,000 + 15000\ P_1$$

If the phase 2 duration is P_2 days, the resultant cost for this phase is given by:

$$\text{Cost}_{\text{Phase 2}} = 3,000,000 + 14000\ P_2$$

If the phase 3 duration is P_3 days, the resultant cost for this phase is given by:

$$\text{Cost}_{\text{Phase 3}} = 4,000,000 + 16000\ P_3$$

Hence, the addition of cost of these three phases results into the cost of entire project as:

$$\text{Cost}_{\text{Total}} = \text{Cost}_{\text{Phase 1}} + \text{Cost}_{\text{Phase 2}} + \text{Cost}_{\text{Phase 3}}$$

$$\begin{aligned}\text{Cost}_{\text{Total}} &= 2,000,000 + 15,000P_1 + 3,000,000 + 14,000P_2 \\ &\quad + 4,000,000 + 16,000P_3 = 15,000P_1 \\ &\quad + 100P_2 + 16,000P_3 + 9,000,000\end{aligned}$$

As the profit to be ensured is 10% of the total cost, the final price will be

$$\text{PRICE} \geq \text{Cost}_{\text{Total}} + 10\%\ \text{Cost}_{\text{Total}} = 1.1\ \text{Cost}_{\text{Total}}$$

$$\text{PRICE} \geq 16,500\ P_1 + 15,400\ P_2 + P_3\ 17,600 + 9,900,000$$

Example 5

Consider the example of law of force and acceleration, i.e., Newton's second law [5]. This law establishes the relation between acceleration and mass of the object. There are many examples in the real world where the Newton's second law is used and applied. Acceleration of the rocket, falling parachutist problem or any other object falling from certain height are some examples of this law. Any mathematical expression is the combination of simple algebraic component to the more complex differential equations. Similarly, the Newton's law states that the rate of change of momentum of object is equal to the resultant force upon it. This law is represented by the well-known equation given below:

$$F = m * a$$

where

F = Total force acting on the body (N or kg)
m = mass of the object (kg)
a = acceleration (m/s^2).

The same equation can be re-written as:

$$a = F/m$$

where a now represents the dependent variable presenting the behavior of system, F represents the force, and m is parameter for property of the system. This equation also depicts the natural process of entire system in more crisp and precise form. Idealization and simplicity of the real-world problems can be easily realized by the mathematical models like this. The use of such mathematical models is multifold which includes precise representation, prediction purpose, proof of concepts, etc. In nutshell, for mathematical modeling, it is very important to understand the domain or field of the problem, what is the entity in the field for which we are targeting mathematical model, organizing principle of the underlined entity and finally, its equivalent mathematical equation. Having known the domain, deciding the entity is the first step. For example mathematical modeling of the problem in data science domain, the data is the main entity for consideration. For mechanical engineering on the shop floor, different mechanical machines are the main entities under consideration. Table 1.1 depicts such parameter matrix required for mathematical modeling in different domains.

For mathematical modeling of engineering problems, practical issues related to engineering applications are very important. Functional aspects and the type of system/problem play crucial role in the modeling process. Linear problems and application of analytical solution for mathematical modeling are quite easy; however exploring nonlinear problems is more interesting and provides more insights and ways. Due to advancements in computer manufacturing paradigm and transformations in algorithms, programming and numerical methods, dealing with the scale of problem has become more easy and feasible. Mathematical modeling of multicomponent system and solving them has become more convenient due to advancements in high computing platforms. Ideal and non-ideal nature, sensitivity analysis and design are also key components of the mathematical modeling. In emerging artificial intelligence and data science applications, data is the main oil for every use case. For unleashing the power of data and demystifying it, reliability test on the underline data set is crucial step before we build the mathematical model. Following are some examples and use cases which will be helpful to readers for initiating the thought process of mathematical model building.

1. Design of mathematical model for the queue management in any mall
2. Resource planning and optimization in an inventory
3. Mathematical model for prediction of cost for real estate project
4. Forecasting and prediction for any pandemic
5. Use Newton's second law to design mathematical model for shock absorber system of the racing car
6. Model to determine the annual revenue model cinema hall
7. Mathematical model to maximize the profit from strawberry cell during pandemic period
8. Mathematical model for entropy in underwater wireless sensor networks
9. Model to optimize the delay in computer and communication network
10. Mathematical model for trust management in ubiquitous computing

1.3 General Steps

Mathematical modeling is very specific to the underlined problem, its type, input and the output. It is very important to understand the nature of problem to decide the appropriate tool or language from mathematics for modeling. In engineering problem domain, there are different types of problem like decision problems, optimization problem and problems with uncertainty, and each type of this problem requires specific mathematical tool and language and is elaborated in Chapter 2 of this book. Solution always resides in the problem. Mathematical modeling helps reader to better understand the problem. Representing problem using mathematics makes clearer understanding due to preciseness of the mathematics and it also helps to reduce the ambiguities [6, 7]. Modeling also enable thorough understanding of the entire system and also gives precision and better direction toward problem solving. General steps in mathematical modeling are depicted in Fig. 1.1.

Step 1: Reducing the Problem

The real-world problem is always complex, imprecise and varied, and first step in mathematical modeling is to reduce the problem. Problem identification by deciding the scope of the problem is crucial step and it helps to draw the clear boundary between what forms the part of the system and what is outside, that is, part of the environment. Once the scope is defined with exact boundaries, making precise assumptions is the next step. It helps later to convert the mathematical model to algorithms and computer program. Assumptions have to be stated very clearly and concisely by exploring various use cases for given problem. Every use case has few ideal situations and these situations need to pay attention as they are easy to handle and help for generalization. There are also some exceptional scenarios which also need to be approximated to ideal situations for the purpose of generalization. Always the set of assumptions varies with the thought process, and in the sequel, for given large and complex problem, several models are possible. However, instead of coming up with the comprehensive model, the simple model is always preferred. Modeling

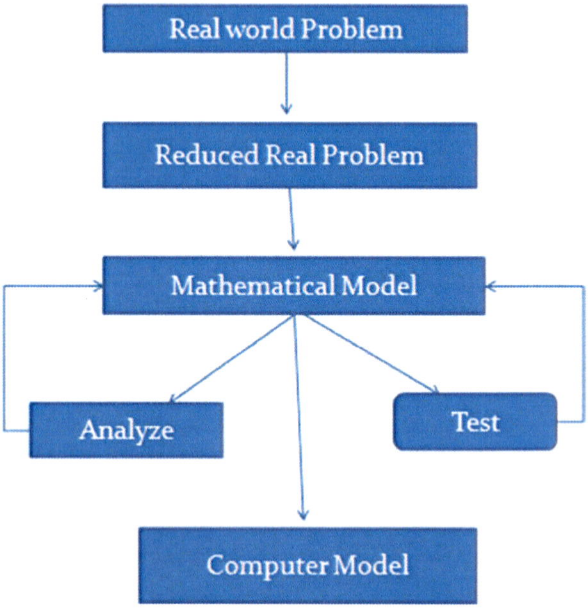

Fig. 1.1 General steps in mathematical modeling

concerns retaining just enough features to make the model useful but not so many as to make it more expensive to compute than necessary to get out the desirable information.

Step 2: Constructing the Model

Every model needs to be tractable, and hence, simplifying it till it becomes tractable is also the crucial step. All the real entities and processes in the entire system are then to be replaced by appropriate symbols, relations and operations from mathematics. The selection of appropriate language of mathematics is the next step in constructing the mode. There may be several mathematical models for the same real situation. There may not be a 'best' model. One can be 'better' than the other in certain situations.

Step 3: Analyzing the Model

Once the problem is formulated into a mathematical form, a rich set of tools can be applied to solve the problem. Several tried and tested methods and techniques are available. The results can be further used to improve the model.

Step 4: Testing the Model

Model testing is next important step in the mathematical modeling which determines whether the model gives accurate predictions. It also verifies whether the predictions are consistent with observations. There are several modeling methods available, and important task is to understand the available tools before we make a choice. Mastering the available tools before we use them enables effective model building.

1.4 Trends in Teaching and Learning

In teaching and learning process, mathematical modeling of the given problem and then proving that problem using proof of concept or other suitable method are of prime importance. Essentially, in the context of research and development, the mathematical modeling of the underlined research problem statement plays vital role in the entire research journey. Mathematical modeling is also preferred and creates more impact in the research publication and math model for a given problem helps to add research component in the contribution. However, there are a lot of myths and realities regarding construction of the mathematical models for the given engineering problem. Many of the times what exactly one has to do in the mathematical modeling is not at all clear to the researchers and it leads to even not a model nor the expression. Merely representing input and output with the help of set theory is considered as a mathematical model. It should be clearly understood that the mathematical modeling requires some concrete philosophy and also requires fundamental knowledge of mathematical tools.

There are many questions which are generally being raised during the process of building the mathematical models and these obvious questions are listed below:

- What is exactly the mathematical modeling?
- Is there any thumb rule for modeling a given problem?
- What skills do we require for mathematical modeling?
- Are multiple models for a given problem possible?
- How to decide which tool of mathematics is more relevant to the given problem?
- If yes, then how to compare these different models for a given problem?
- How to analyze and test the mathematical model?
- Whether the mathematical modeling is applicable to the problem or solution or to the entire process of problem solving?
- How to convert mathematical model in to the computer program?
- Which are the performance evaluation parameters of mathematical model?

Discussion on these questions and possible solutions is presented and discussed in the scope of next chapters.

1.5 Summary

This chapter has extracted and presented the fundamental concept of modeling in the real world with examples. This will provide an understanding to the interested readers what is exactly the modeling. The need of modeling with reference to the real-world use cases is also presented and discussed in this chapter. The examples and case studies to show the applications of the all the explained concepts are a bonus that will help researchers in understanding the concepts better and find a connection to their problem. In the next section of this chapter, the mathematical modeling is elaborated by taking more examples so that reader will have clearer idea of math modeling. General steps of mathematical modeling are depicted further with the help of an introductory approach. This chapter finally concludes with the myths and realities of mathematical modeling and trends in teaching and learning.

References

1. Dromey RG (2005) How to solve it by computers. Prentice Hall
2. Miller TW (2018) Modeling techniques in predictive analytics with python and R: a guide to data science. Pearson Education
3. Mityushev V, Nawalaniec W, Rylko N (2021) Introduction to mathematical modeling and computer simulations. Chapman and Hall/CRC
4. Garrido J (2011) Introduction to elementary computational modeling essential concepts, principles, and problem solving. Chapman and Hall/CRC
5. Knight RD (2008) Physics for scientists and engineers: a strategic approach, 2nd edn. Addison-Wesley. ISBN 978-0805327366
6. Dangelmayr G, Kirby M Mathematical modeling a comprehensive introduction
7. Neumaier A Mathematical Modeling

Chapter 2
Problem Solving and Mathematical Modeling

2.1 Problem Evolution

A problem is a puzzle or task that requires a logical thought process or fundamental mathematical steps to solve it. The puzzle generally represents the set of questions on the underlined use case which also consist of complete description of the use case along with the set of constraints about the use case. Logic is very importantly used to solve the puzzle or problem. Logic is defined as a method of human thought that involves thinking in a linear, step-by-step manner about how a problem can be solved. Logic is a subjective matter and varies from person to person. Logic is directly linked to the natural intelligence of human being and is a language of reasoning. Logic also represents the set of rules we use when we do reasoning. It is observed that majority of the employment of fresh engineering graduates across the globe is in software and information technology sector.

As per the report [1, 2], over 80% of the engineering graduates prefer to take up immediate job rather than going for higher education essentially in the Indian context. However, it is also noticed that only 20% of these fresh graduates are industry ready and remaining required need-based training on the contemporary skills in order to place them on the live projects. The basic skills expected by information technology employers are problem-solving skills, quick learning ability, applied knowledge and soft skills. As per [2], the basic problem-solving skill in the respective domain is mandatory requirement which is missing in the fresh engineering graduates and they require training on the same. In the view of this, it is important to understand and revisit essential skills needed for a software engineer to demonstrate the need for logical thinking for problem solving. For developing good problem-solving skills, it is very important to understand the given problem first before we get into the solution domain.

P. N. Mahalle et al., *Foundations of Mathematical Modelling for Engineering Problem Solving*, Studies in Autonomic, Data-driven and Industrial Computing,
https://doi.org/10.1007/978-981-19-8828-8_2

To simplify it further consider the following examples:

Example 1 Scenario: VIIT campus has a library. Librarian issues books to only VIIT students and staff.

Careful observation suggests....

Description of reality: There is a VIIT library in campus and there is a team of library staff.

Constraint: Librarian issues books only to VIIT students and staff

Questions about the use case:

1. How many books are there in the library?
2. How many books can be issued to the staff?
3. How many books can be issued to the student?
4. For how many days books can be issued to the staff/student?
5. Does the librarian issue books to himself/herself?

Example 2 Scenario: The objective is to put clock on the wall.

Careful observation suggests.

Description of reality: The clock is given and the problem is to put this clock on the wall at appropriate place using available resources.

Constraint: Use available resources to place the clock on wall.

Questions about the use case:

1. At what height?
2. Which Wall?
3. Where on the wall?
4. By what time it should be done?
5. Should it be nailed or pasted?

Above two examples show that the answers to the questions generated in above two examples lead to proper understanding of the problem, and also, it results in preparing a list of tasks to be performed. It results in preparing a list of tools or resources required to solve the problem. In the current trend, most of the problems are solved with the help of computers by developing programs from algorithm through appropriate programming language. However, it is very important to solve the same problems on paper before we solve them using computer. Representing the information which is available to describe problem in desired form and then finalizing the steps to transform this information from one representation to another are two main steps before we get into problem solving. Creative approach for engineering problem solving requires exploring the powerful tools for solving the complex problems. There are many times the contradictory requirements for the given problem and it requires capability of making problem self-elimination. There should be proper mechanism to deal with the sub-problems of given problem for implementation solution. The theory of constraints (ToC) plays crucial role in the engineering problem solving and modeling [4, 5]. The important and limiting factors which are crucial to consider in the problem solving come from the ToC. ToC is one of the important problem-solving methodology which helps to identify the constraints and later on how these constraints can be incorporated in the solution. The more detailed methodology of problem solving is presented and discussed in the next section.

2.2 Problem Solving

When it comes to problem solving in the field of engineering, mathematical modeling becomes an integral part of the project that helps in expressing the efficacy and the real-time applicability of the solution. With the mathematical modeling forming the foundation of the research work/project, there arises a necessity for the researchers to be able to understand and connect the mathematical concepts with that of the underlying research problem. The main objective of this section is to discuss the evolution of a problem and the concept of problem solving along with the problem classification aspect which is a key aspect in mathematical modeling process. Understanding the problem classification concept makes the modeling process easier. It is also important to understand the modeling process and the teaching approaches that are followed principally.

The engineering problem solving is a very structured process where the first step is to define and formulate the given problem with the set of questions as well as the set of constraints to be associated in solution. Let's revisit the example 2 presented in Sect. 2.1. While evaluating the questions 1–5, it can be seen that few of the questions will be converted to constraints which need to be incorporated in the solution. In Example 2, where the objective is to put the clock on the wall, questions like at what height and which wall will become the constraints in the solution phase and they need to incorporate in the problem-solving process. In the view of this, the reformulated problem which considers these constraints is as follows:

Reformulated Problem:

"The problem is to put 30 grams of clock on the left wall in center at 12 feet height with the help of nail"

It is very clear that the reformulated problem will help to reach to the solution more quickly as compared to the earlier problem. As the new problem description consists of the detailed description of reality with set of questions and set of constraints to be incorporated in the solution, i.e., ToC, it leads to the quicker and efficient problem solving.

The process of problem solving has multiple aspects and dimensions, and also, the type of problem which needs to be solved plays important role in the entire process. The detailed workflow of problem solving and its evolution is depicted in

Fig. 2.1. The process of problem solving is at the root and has two sub-trees: one branching to the problem domain and another branching to the solution domain. The problem under consideration belongs to one of the several categories and it is important to decide the category before we get into the solution domain. Every problem needs solution and it can be achieved with the help of algorithm which is nothing but the representation of computation. Flowchart or pseudo code is used to represent algorithm in engineering domain as it is easy to convert them into the computer program. Algorithms also need to meet the requirements which include finiteness, definiteness, effectiveness and defined input/output. The category of the algorithm under design also plays crucial role in developing the better model. The entire problem-solving process along with their types is elaborated below:

1. **Problem Categories**
   a. **Sequential**—Operations are performed in step-by-step manner in sequential category. There is a dependency for the sequence of operations to be carried out.
   b. **Even based**—In event-based category, the operations are performed based on some inputs. During execution, when the value of variable or some condition matches, the business logic is executed to take some action.
   c. **Concurrent**—In concurrent category of the problem, operations overlap in time. There is no dependency for the sequence of operations to be carried out.
   d. **Distributed**—In distributed problem, the operations are performed at different locations to perform the given task.
2. **Algorithm Patterns**
   a. **Sequential**—This pattern executes the statements in the order in which they appear in the algorithm. Sequential execution of the statements is the key feature of this pattern.
   b. **Selectional**—This pattern controls the flow of statement execution based on some predefined condition. The remaining part of the code within the scope is executed based on this condition.
   c. **Iterative**—This pattern is used when the particular part of algorithm needs to be executed multiple times.
   d. **Recursive**—When the underlined problem requires the functions computed by algorithms need to be expressed in terms of itself, then the recursive pattern is used. This pattern is easy to model, however computationally expensive in implementation.

Furthermore, problem categories and algorithmic pattern also decide the appropriate tool from mathematics to be used for modeling. As these two parts play important role in formulating the solution in algorithmic perspective, analysis of given problem in these aspects eventually leads to the success of modeling. This analysis also helps to decide the appropriate data structure, which is required in problem-solving process.

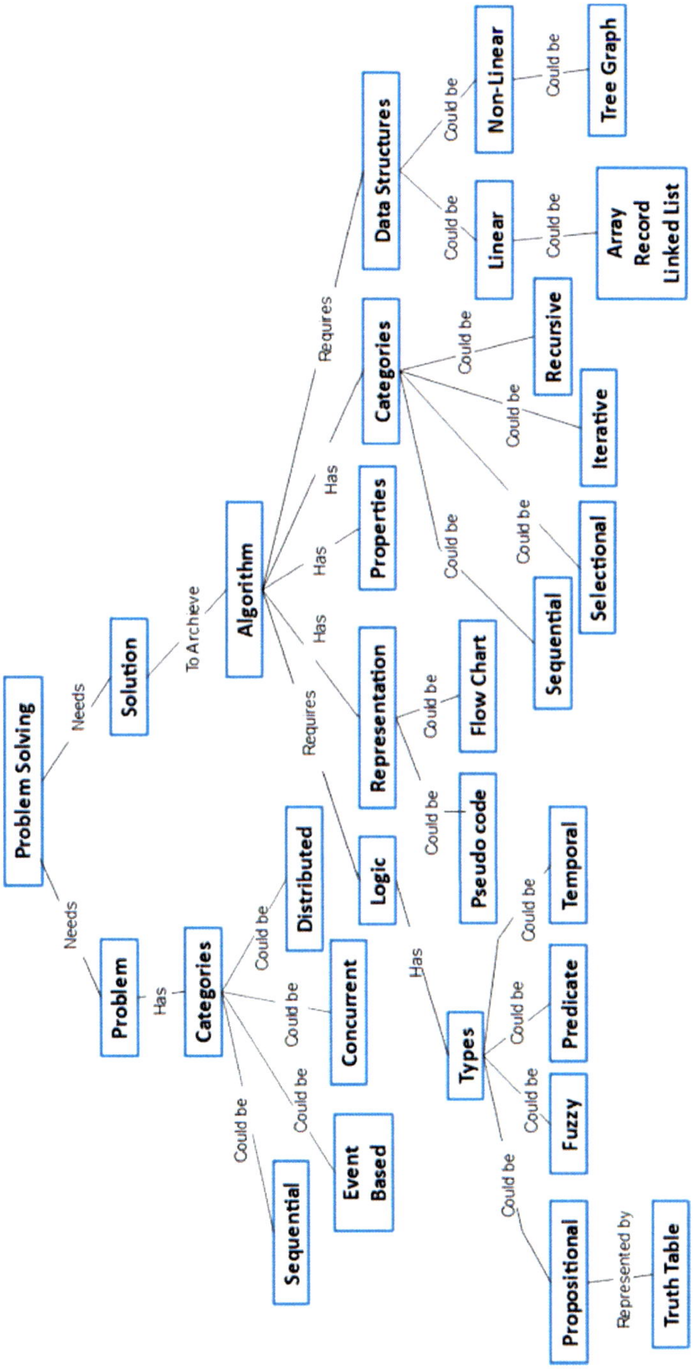

Fig. 2.1 Problem solving

Pólya and George, the famous mathematics educators [6], believed that the skill of problem solving could and should be taught—it is not something that you are born with. As stated by him, there are four founding principles of problem solving, which are stated below:

1. **Understand the problem**—What is the problem? If you do not know what it is, how can you solve it? The main principle is to understand the problem fundamentally.
2. **Devise a plan**—Design a strategy, there are many known strategies such as guess and check, or solve a simple problem or divide problem into sub-problems. Use one of them or come up with a new strategy or action plan. This principle states that one should use appropriate algorithmic strategy to solve the problem.
3. **Carry out the plan**—Execute or implement the plan; be patient—most problems are not solved quickly nor on the first attempt. If a plan does not work immediately, be persistent. Do not let yourself get discouraged. If one strategy is not working, try a different one. This principle states that planning plays crucial role in problem-solving phase.
4. **Look back (reflect)**—Is there a better way? Have I solved it completely or there is still need for refinements. Ensuring whether the problem is completely solved or change management is require to solve it further is the main objective of this founding principle of problem solving.

Problem understanding requires stating the problem by elaborating it or drawing a picture or sketch. Written statement helps in focusing on the main objective of the problem, and also, it avoids losing the track. One should not try to solve the problem before it is described to yourself in precise manner. Many times in a hurry to solve the problem, one gets started without understanding what the problem is and eventually end up solving a wrong problem. There are always many ways of solving a problem and it is difficult to decide where to start. It is recommended to use some tried and tested methods of problem solving. Proven strategies that had given fruitful solutions earlier can be used, and these strategies also help in rearranging the tasks in proper sequence, associating tools with tasks. Your choice of strategy may lead to knowledge of new tools which you had not earlier envisaged. Your knowledge of tools can also affect the strategy you tend to choose. Implementation requires good mastery over the tools to be used in solving the problem. Gathering the knowledge of tools is essential for problem solving, but this does not make you a problem solver. Sometimes we have the action plan ready and know the tools required for it but are unable to use them. We need to master tools in anticipation that we may need them in some futuristic problem-solving effort. Skilled problem solver will get it right in the first attempt is a myth. It is usually based on the above steps critically. The set of general questions is as follows:

- Whether the problem is defined properly?
- Is the problem specification complete?
- Does the solution only partly solves the problem or only tackles specific cases or is not taking care of all possibilities?

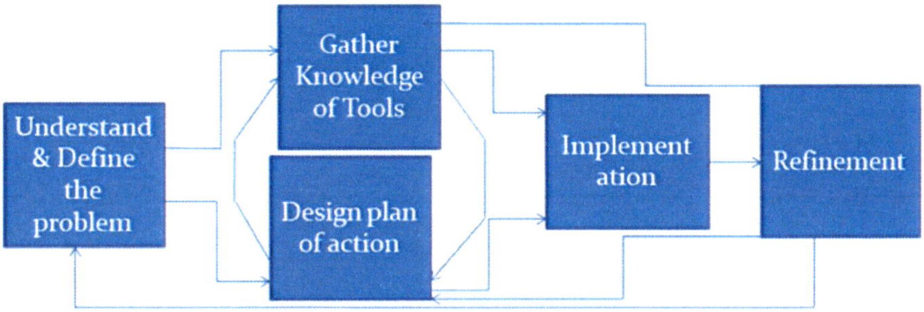

Fig. 2.2 Problem-solving phases

- Are there better tools?
- Do the tools have better capabilities?
- Using the new tools the quality of solution could be improved?
- Are there better designs possible for the problem? Better in terms of cost saving.
- Will the new action plan result in faster, efficient good quality solution?
- Is the implementation done properly?
- Are there bugs, errors?

The mathematical modeling generally creates the mental block, and to overcome it, one jumps to the next phase of implementing the solution without preparing the action plan. Mathematics provides a rich source of tools for problem solving and it would be advisable to use such tried and tested methods. However, mathematics can only be applied to mathematical problems and mathematical modeling is a way of formulating a given problem in a mathematical form. These problem-solving phases are depicted in Fig. 2.2.

2.3 Problem Classification

Analyzing and classifying the problem is a central theme of problem solving. Essentially, looking into the skills require for software engineers includes technical skills like software designing, coding and software testing, problem-solving skills like logical and analytical thinking, problem modeling and the soft skills like communication, teamwork and negotiation. In order to develop the problem-solving skills, logical and analytical skills are the main pillars. Logical thinking enables thinking in terms of cause and consequences in sequential manner initiating the train of thoughts. Logical thinking also helps in decision making, problem solving as well as mathematical modeling of the problem. In the sequel, logic also varies from problem to problem and the specific types include propositional logic, predicate logic, fuzzy logic and tense logic. However, the propositional logic is most useful for engineering problem solving to develop solutions to the problem [7, 8]. In daily life, we come across different types of problems like social problem, managerial problems, computational problems, etc. In engineering domain, most of the problems are computational problems which involve the transition from one state to another state with some predefined rules and set of computations. The different types of computational problems are listed below:

1. **Decision problems**

   These are the type of computational problems in engineering domain; the answer to every instance of the problem is either yes or no. Deciding whether a given problem is polynomial time complexity is the example of decision problem.

   A decision problem is a query in a formal system that has a yes-or-no answer based on the values of some input parameters, as defined by computability theory and computational complexity theory. Motivation/scenario is very important that presents a simple decision problem to the readers for better understanding. The research problems/projects as case studies play important role to explain the application of mathematical modeling in real time.
2. **Optimization Problems**

   While there is only one correct solution for each input in decision problems, optimization problems are focused with finding the best answer to a certain input. Finding the best solution out of several feasible solutions to either minimize or maximize, given optimization function is other type of computational problem and is referred as optimization problem. Maximizing the production of items on the production floor is an example of optimization problem.
3. **Delay Model Problems**

   The problems with formation of queues results into the delay model problems. Delay is entrenched in many natural and man-made systems, including biology, process industries and mechatronic motions. Time delays are used to characterize propagation processes, material or energy transfer in interconnected systems, and data transmission in communication networks in engineering applications. They are the primary causes of oscillations, instability and poor control performance. Therefore, the stability analysis and robust control of such systems have gained theoretical and practical importance in research and development.
4. **Data Science Problems**

   A data science problem is focused on deriving conclusions from the available data. A true data science problem focuses on providing one of the following as output, namely categorizing or grouping of data, identifying patterns, identifying anomalies, showing correlations and predicting outcomes. One of the key skills required in modeling a data science problem is the ability to precisely evaluate and define the problem at hand.
5. **Pandemic Problems**

   Pandemics are large-scale epidemics of infectious disease that can cause major economic, social and political upheaval while also increasing morbidity and mortality across a huge geographic area. Epidemiologists model the mechanics of disease spread in order to stop an outbreak. During a pandemic, epidemiologists must also consider the likelihood of hospital beds becoming overcrowded, the disease spreading to other areas and caregivers running out of personal protective equipment and so on. Hence, mathematical models have become invaluable in such crisis situations.

6. **Interdisciplinary Engineering Problems**

   The increasing global demand for novel mathematical answers to complicated problems has led to a greater understanding of the value of interdisciplinary research in the mathematical sciences and with other fields. A strong base in the mathematical sciences and modeling approaches among the students and researchers is required to be able to tackle difficult issues, adding intellectual value to emerging technologies, leading innovation and remaining globally competitive.

2.4 Modeling Process and Teaching Approaches

Mathematical modeling in a nutshell is the representation of a real-life problem in an appropriate mathematical form. Mathematical models are used across various disciplines such as engineering, natural sciences, social sciences and even in music, philosophy and religion. The analytical basis for the design and control of a system can be well established using mathematical modeling in the field of engineering and technology. This eventually helps in making reliable predictions, thereby minimizing the expenditure of effort, time, money and other valuable resources. Mathematical modeling requires the set of competencies which include identification of questions, targeted variables, equivalent functions or assumptions related to the real-world entities, translations of these components into the mathematics, solution validations and analysis/comparison of the solution with state of the art.

As stated in [9], there are broadly three categories of the mathematical modeling as listed below:

- ***Intra Mathematical Modeling***—it uses linear algebra and related mathematical tools to solve geometrical problems.
- ***Direct modeling***—it is used to solve traditional word problems.
- ***Complex modeling***—it is used to solve different types of open problems from the real world.

The complex modeling is very important and its application in interdisciplinary areas is considerably increasing in recent days due to the transformations in research. Consider the following examples to get more insights into these types of problems which are depicted in Table 2.1. Table 2.1 presents and discusses the examples of real-world problem and their type.

The modeling problems come from the real world which is generally formulated in everyday knowledge. First step is to understand the task in order to make a mental

Table 2.1 Common types of problems

Sr. No	Problem	Remarks
1	The problem is to find out different possible menus which can be offered from the menu card of restaurant. One combo includes starter, main course and dessert. The available items are four varieties of starters, five main courses and three desserts	This is an example of combinatory problem
2	The problem is to help milk supplier company to reduce the cost of milk can by reducing the cost of material required for fabricating the can. Milk supplier wants us to develop new cylindrical milk can, which will contain 0.5 L of milk	This is an example of optimization problem with the given constraints
3	How long will it take to evacuate the main building of college which is under fire?	This is an example of forecasting or prediction problem which involves uncertainty

representation of the given task. This helps to percept about individual understanding of the problem and task formulation. The type of the problem and domain helps to perform systematization of the problem, and then, next step is to perform mathematization to build the mathematical system. In the sequel, this mathematical system is analyzed to produce the model results, and then, the interpretation and evaluation of the model are carried out to validate it. This modeling cycle is depicted in Fig. 2.3.

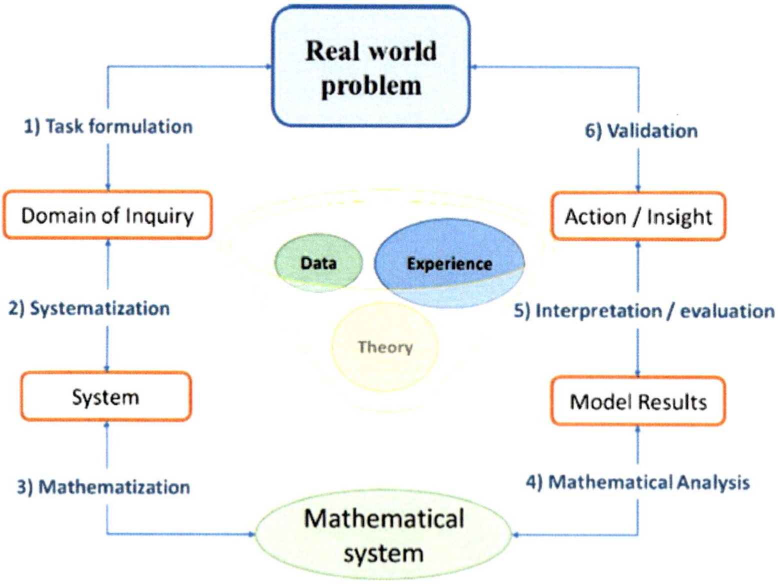

Fig. 2.3 Modeling cycle

Table 2.2 Myths and realities in mathematical modeling

Sr. No	Myths	Realities
1	It is difficult to understand	It is easy and uniform to understand
2	There is ambiguity to interpret	It is always unambiguous and clear to understand
3	Mathematical models are cross-cultural	Mathematical models are not cross-cultural
4	Linguistics differences are not the barriers in mathematical modeling	Linguistics differences are the key barriers in mathematical modeling
5	Modeling is applicable to only problem	Modeling is applicable to both underlined problem and solution

Teaching Approaches

Mathematical modeling is considered as one of the tough tasks in the process of research and development. In fact, the modeling process of the problem using mathematics is hard-hitting process if the problem is not understood clearly and the modeling process is not clearly defined depending on the type of the problem. In teaching fraternity, unfortunately the mathematical modeling is considered as just representing the problem using set theory, functions or relations. However, it is beyond it and it requires the problem to be classified into the appropriate category and then selecting the right tool from mathematics for right problem. Interesting insights into mathematical modeling in teaching are presented and discussed in [10, 11]. Trends in teaching and learning for mathematical modeling are described in [12]. The competencies required for real-world mathematical modeling, aims and perspectives of modeling, the teacher support required for modeling activities as well as the strategies for learning the modeling process is presented and discussed in [10]. Before we get into the modeling part of different categories of the problem, it is important to understand the myths and realities regarding mathematical modeling. These myths and realities are presented in Table 2.2.

Mathematical model is considered as difficult to understand; however, it is always easy and uniform to understand if the reader is having fundamental knowledge of mathematics and the given problem. If the problem is represented in English, then it is likely to be interpreted in different ways by different reader. If the same problem is represented with mathematical model, then there is no ambiguity in its interpretation and it is always clear and uniform to understand. Generally, it is considered that the mathematical models are cross-cultural in technical community. Mathematics is a language of representation and problem solving; algorithms are the main pillar of this language. Algorithms are having different representation and they are not uniform across various cultures and nations [13]. In addition to this, different mathematical symbols are used for different purposes. Across cultures and nations, there is diversity in the notations, symbols and processes. Linguistic differences also play important role in the mathematical modeling. Being English as a business language, many researchers and students not having English as a primary language face lots of issues in building the mathematical models. In the sequel, it is recommended to consider these myths and realities in building the mathematical models of the given problem.

2.5 Summary

Problem solving plays vital role in the process of mathematical modeling. Due to increasing volume of research in science and technology, mathematical modeling and its application in the research is transforming across the globe. This chapter begins with a briefing about the evolution of a problem and the concept of problem solving. Later, the chapter discusses about the problem classification aspect, which is a key aspect in mathematical modeling process. Understanding the problem classification concept makes the modeling process easier. Finally, the chapter elaborates the modeling process and the teaching approaches that are followed principally.

References

1. Rao ORS Hiring trends of fresh engineering graduates. Report in engineering special
2. https://www.linkedin.com/pulse/story-fresh-engineering-graduate-indrani-mukherjee/
3. Joshi SA, Mahalle PN (2016) Design and analysis of algorithms: a problem solving approach. Cambridge University Press, India, p 209
4. TOCICO Board of Directors—Theory of constraints international certification organization. www.tocico.org. Retrieved 17 Mar 2018
5. Dettmer HW (2007) The logical thinking process: a systems approach to complex problem solving. ISBN: 978-0-87389-723-5
6. Pólya G (1945) How to solve it. Princeton University Press. ISBN: 0-691-08097-6
7. Dormy RG (2005) How to solve it by computers. Prentice Hall
8. Levitin A (2005) Introduction to design and analysis of algorithms. Pearson Education Press
9. Frejd P (2014) Modes of mathematical modelling: an analysis of how modelling is used and interpreted in and out of school settings. Linköping University
10. Blum W (2015) Quality teaching of mathematical modelling: what do we know, what can we do? In: Cho S (eds) The proceedings of the 12th international congress on mathematical education. Springer, Cham. https://doi.org/10.1007/978-3-319-12688-3_9
11. Ball DL, Hill HC, Bass H (2005) Knowing mathematics for teaching. Am Educ 29(3):14–46
12. Biccard P, Wessels DCJ (2011) Documenting the development of modelling competencies of grade 7 students. In: Kaiser G et al (eds) Trends in teaching and learning of mathematical modelling (ICTMA 14). Springer, Dordrecht, pp 375–383
13. Kantner J (2008) The only absolute truth in mathematics is the myth of mathematics as universal. Perspect: N Y J Adult Learn 6(2). Retrieved from http://www.eric.ed.gov/PDFS/ED501486.pdf

Chapter 3
Decision Problems

3.1 Introduction

The problem which takes inputs (like graph G, Tree T, Matrix M, etc.) and gives the output 0/1 or Y/N is known as decision problems. In algorithmic, decision problem covers wide range of problems in computing. Some popular decision problems in daily life are whether in the city, all the squares are reachable from each other, i.e., all the squares in the city are connected or not connected, can we reach from square 2 to square 7 in less than 5 km distance? Whether the given graph is bipartite or not? This shows that, in other category of the problems, outcome is some number or significant output and in decision problem, the output is a binary value (0/1—yes or no).

Consider the scenario where we are given with an empty knapsack with some capacity M and n objects are given with their weights and profit. For this instance, can we get a profit of at least k is referred as a decision problem and it takes polynomial time to verify that whether we can get the profit of at least k. Likewise, various mathematical areas such as probability and statistics, optimization, queuing theory, control and game theory have been used to model the decision-making process mathematically, thereby allowing qualitative outcomes. Topics such as optimization, queuing theory, probability and statistics have been discussed in Chapters 4, 5 and 6, respectively. In this chapter, we elaborate upon control theory, game theory, decision trees and multicriteria decision analysis and its use in decision-making process. We further use a few case studies that explain the application of mathematics in decision-making process.

3.1.1 Control Theory

Control theory [1] is associated with the management of control in dynamic systems. The theory focusses on achieving a certain degree of optimality in the process or

P. N. Mahalle et al., *Foundations of Mathematical Modelling for Engineering Problem Solving*, Studies in Autonomic, Data-driven and Industrial Computing,
https://doi.org/10.1007/978-981-19-8828-8_3

machine. Control theory is used to progress the system to a desirable state by developing an algorithm or model that governs the application of inputs to the system. The algorithm efficiently minimizes the delay and guarantees stability in the system. An important component used in this process is the controller which tries to bring the controlled process variable (PV) to the desired reference value, i.e., the set point (SP). The difference between PV and SP given by SP-PV is calculated to be the error signal and is feedback to the system repeatedly to achieve the desired result. Controllability and observability are two other aspects that are the basis of the advanced automated systems including aircrafts and communications in industries. The sensors obtain measurements after which constant monitoring and adjustments to the measured values are made to keep them within the set limit with the help of the final control element. Block diagrams are extensively used in control theory. Here, the transfer function mathematically represents the relation between the input and output in line with the differential equations that mathematically describe the system.

Typically, there are two types of control loops, namely open loop and closed loop. In the open loop systems, there is no connection between the desired value or set point (SP) and the controlled variable or process variable (PV). For example, consider a central heating system controlled by means of a timer. The heat is applied for set time despite the temperature in the building and then cut off when time out is reached. Here, the temperature of the building is the process variable which is never measured or used in controlling the temperature of the system.

In closed loop systems, the feedback element which is calculated as SP-PV is monitored and is feedback to the system to achieve the desired result. For example, in the cruise control of a car, the desired speed or SP is set by the driver. The cruise control acts as the controller. The car and the cruise control form the entire system. The system output or the PV is the car's speed. The engine's throttle position determines how much power is to be applied to the system in order to maintain the speed. When there is an increase or decrease in speed due to a downhill or uphill encounter, the difference between the system's current speed (PV) and the set speed by the driver (SP) is calculated and is feedback to the system which determines the throttle position to direct the amount of power to be applied by the engine.

A typical mathematical representation of the closed loop transfer function is shown in Fig. 3.1 with the help of a block diagram.

Here, the output of the system is represented as $y(t)$ which is fed back into the system via F which is the sensor measurement. It is compared against the reference value or set point given by $r(t)$. The difference obtained is given by e and is fed to the controller C, which alters the inputs u to the system according to the control P. Using Laplace transform to analyze the system by considering the components C, P and F to be linear and time invariant, we get,

$$Y(s) = P(s)U(s)$$

$$U(s) = C(s)E(s)$$

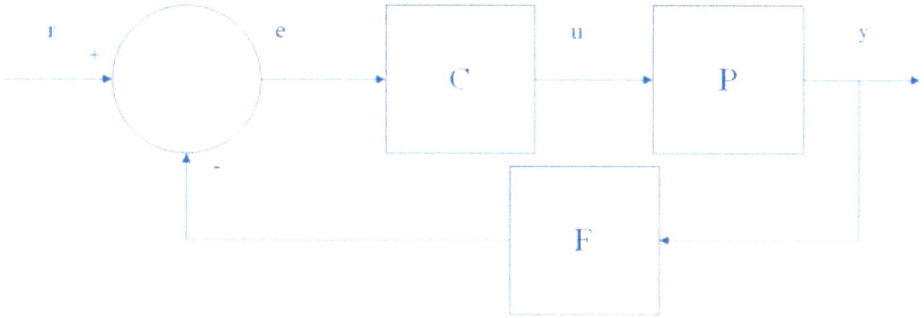

Fig. 3.1 Closed control loop block diagram

$$E(s) = R(s) - F(s)Y(s)$$

Solving for $Y(s)$ in terms of $R(s)$ we get,

$$Y(s) = \left(\frac{P(s)C(s)}{1 + P(s)C(s)F(s)}\right)R(s) = H(s)R(s)$$

Here, $H(s) = \frac{P(s)C(s)}{1+P(s)C(s)F(s)}$ is the closed loop control function. The numerator is referred to the as the forward gain from r to y and the denominator is loop gain.

3.1.2 *Game Theory*

Game theory [2] is a field that finds its application in adaptive systems that change their behavior according to changing circumstances. The individual elements in the system are adaptive and interdependent. Game theory assists in understanding and mathematically modeling such adaptive systems. Choices by agents in the adaptive systems that affect their outcomes are the major driving forces in game theory. Games are formed out of the interdependencies between adaptive agents. An agent is any entity that has an agency. Agency refers to the ability of the agent to make decisions by considering the available information. They can act upon their choices autonomously, thereby affecting the environment's state. The resources that the agents require have a value, referred to as the goal. Understanding an agent's value is important in modeling a game. Agents can also be defined by their goal-oriented behavior. To sum up, a game consists of agents which are built upon a value system. Agents make choices of their own whose actions might affect other agents. The resultant effect of the interactions in the game has a payoff. The routing of Internet traffic and the interaction between algorithms can be modeled as a game. A typical example of the game theory can be the current political state of all nations and the environmental impact. It requires all nations to work cooperatively toward the goal of reducing environmental impact.

However, it comes with a price where nations might have to compromise on industrial productions and growth to reduce the toxic emissions. These decisions must made by individual nations based upon their cost–benefit analysis. This is like the classical example of game theory called the prisoner's dilemma which captures the core dynamics of game theory which are cooperations and competition. The interdependence between agents gives rise to two level in the system, namely the macro-level and micro-level. The macro-level represents the state where the individual agents will have to work cooperatively to achieve the desired goal. At the micro-level, individual agents pursue their own agendas by performing their own cost–benefit analysis. This gives rise to the social dilemma concept where the rules governing the policies and decisions at the macro-level are different from that of the micro-level. What might seem rational in working in a cooperative mechanism might seem irrational at the individual micro-level. Complexity arises due to the existence of different rule sets and trying to align the interest of individuals at the micro-level with that of the group in the macro-cooperative level. All games have three components, namely players, strategies and payoffs.

A player performs the decision making. A specification of each decision in every situation that the player might find himself is called as a strategy. The reward or loss that all players must face because of following its own strategy is called a payoff. Games are generally represented in either a matrix form or a tree graph.

The matrix model is the most common method used to model static games which do not involve time. Players choose their strategies simultaneously in such games. It is also called as the normal form representation. The players are associated with the axis of the matrix. Each column or row depicts a strategy that is unique for the player. When the interaction of players' strategies in the matrix, values are placed that correspond to the associated payoffs for the player when the strategy is executed. The players do not know the decisions of the opponent while making their choices. Figures 3.2 and 3.3 represent the matrix and tree graph models of games respectively.

A tree graph model is used to model dynamic games where time is involved. It allows the players to make choices in a sequential manner along the course of time. The tree graph captures these choices as traces for each player. It is an extensive form. An example of such a game is the chess where the players make decisions based upon the previous players' moves at every instance. The agent makes calculated decisions about where the current choice might lead him to and then takes an informed decision for the current move.

Therefore, it is evident that in any game, the agents make decisions based on the information that they have, and this makes it the second important aspect in the game. Agents can therefore have complete information including the payoff and strategies of the opponents or there can be incomplete information where the other players' payoffs and strategies are unknown. A game in which the payoffs for players for different possible actions are not the same is called as an asymmetrical game. If the players' identities can be changed with no need of altering the associated payoffs, then the game is said to be symmetrical. Games are played over a mutually desired resource associated with a value. There always exists some common understanding of what the agents' value is and how the value is dispersed out depending upon

		PLAYER 2			
		R,L	L,R	R,R	L,L
PLAYER 1	U	5,2	2,5	5,2	2,5
	D	0,0	3,1	3,1	0,0

Fig. 3.2 Matrix representation of games

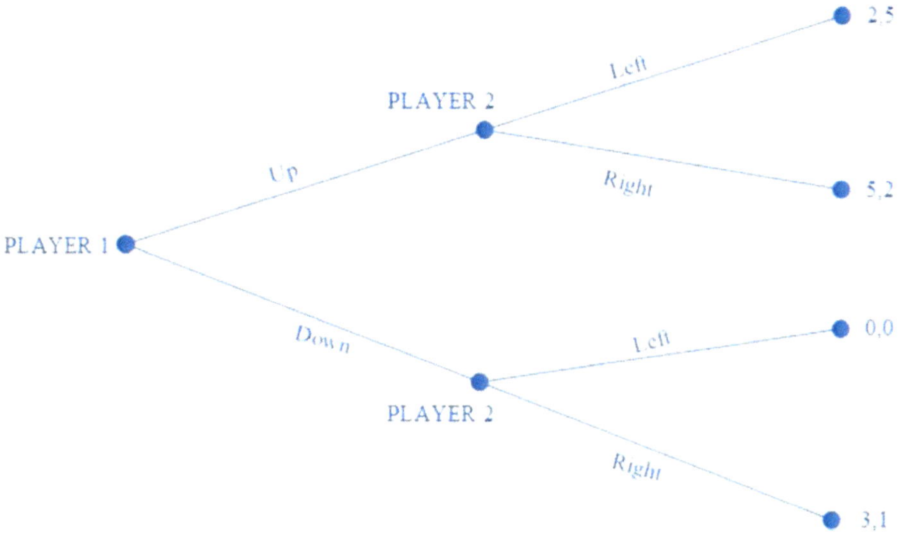

Fig. 3.3 Tree graph representation of games

the agents' actions. However, does the total value distributed among agents remain the same regardless of their actions or can it increase or decrease subject to their capacity to cooperation? In constant sum games, the sums of payoffs of the players add up to be always the same. Here, one person's game is another person's loss. The payoffs can neither be increased nor decreased in a game because of player choices in a special case of constant sum games called the zero-sum games. Here, the total benefit to all players for every combination of strategies always adds to zero. Here, the relationships between agents' payoffs are negatively correlated called the negative interdependence where there is no space or payoff for cooperation, and hence, competition is key in such games. The level of cooperations among the players decides the rise or fall of the total value to be dispersed among them in a non-constant sum games or non-zero-sum games. Here, the outcome of agents is positively correlated, as in businesses where all agents work together for successful outputs. If one agent gets more payoff, then the other agents also get more. Hence, there exists positive interdependence where all players work in cooperation leading to a goal. Cooperative games are an example of non-zero games. Here, either every player wins or loses. Cooperation may be achieved in many ways such as in-built within the game, or imposed by some external factors, or through peer-to-peer interactions and feedback mechanisms. Non-cooperative games hold an element of competition. Here, there are not many possibilities of creating an environment of cooperation because of the game's inherent nature, thereby making it a zero-sum game. The game is strictly competitive and there exists no value for cooperation.

3.1.3 Probability and Statistics

The concept of decision making has improved with the adaptation of mathematics and especially **probability theory** over the years. Decision making in engineering and technology becomes more complicated as it involves situations of risk and uncertainty. The whole idea of using probability in engineering decision making is that the outcomes are always uncertain; however, the choice of assigning or not assigning values to the outcomes depends upon the evaluator and the nature of the problem. In decision making under uncertainty, the outcomes are uncertain, but we do not assign probabilities to the outcomes. In decision making under risk, the outcomes are uncertain, but we assign probabilities to the outcomes. The most used quantitative methods for decision making under risk and uncertainty are as follows.

Single point estimate is a statistical measure which determines the performance measure by using a single estimate of every unknown. It is like using the mean value or averages to make decisions. They can be used for both risk and uncertainty. Here, only one output value is obtained for each alternative. Decision rules need not be applied further as there is only one performance value for each alternative. The alternative with the highest net present value (NPV) can be chosen to improve/maximize the performance measure.

Scenario Analysis is an extension to the single point estimate. Here, unlike the single point estimate, calculations are done many times for each alternative by changing the input variables. It is also applied under uncertainty and risk. For decisions under uncertainty, we look for worst, best and most likely cases. Table 3.1 shows the decision matrix under uncertainty. However, for decisions under risk, we assign probabilities to each outcome of the model and represent the same in the form of a decision matrix as shown in Table 3.2.

Break even analysis is a reverse modeling process where an output is set for the break-even point, and then, we solve for the input variables.

Table 3.1 Decision matrix under uncertainty

	Case 1	Case 2	Case 3
Alternative 1	−200	250	320
Alternative 2	−300	275	310
Alternative 3	−350	260	375

Table 3.2 Decision matrix under risk

	Case 1	Case 2	Case 3
Probability	0.2	0.55	0.25
Alternative 1	−200	250	320
Alternative 2	−300	275	310
Alternative 3	−350	260	375

Sensitivity analysis provides an understanding of the way in which the change in input variables might affect the output of the model. Determining the variables to which the output is highly sensitive will help us in making decisions on variable variation limitation, to monitor and perform adjustments with time, make better estimates and alleviate the risks to variables at a certain cost.

Decision Trees are generally applied to scenarios under risk category. A decision tree has three types of nodes as shown in Fig. 3.4, namely the chance node showing the probabilities associated with the results, represented by a circle; the decision node that indicates any decision to be made, represented by a square; and an end node that typically indicates the outcome of the path taken for the decision is presented in the form of a triangle. Every node originating from a chance node must have probabilities assigned to them. A chance node in general is a representation of the existing uncertainty in the decision path. Decision trees help in calculating the expected value or utility. We do this by starting with the end nodes, working back to the root node while calculating the expected value at each intermediate node.

Monte Carlo Simulation provides [3] a histogram representation of the input values along with their frequencies. Random sampling of input variables is done, and the resulting output is noted. This is repeated for a thousand records resulting in a histogram from which a range of potential outcomes along with their probability of occurrences can be determined.

The following are the decision rules used to evaluate various alternatives under uncertainty in which we depend merely on the outcomes for decision making.

In *Laplace criterion* [4], we assume that every outcome is equally likely events (have equal probabilities of occurrence). In this method, the alternatives are evaluated by taking an average of all the outcomes and the alternative with the highest average is chosen.

Alternative 1:—$(-200 + 250 + 320)/3 = 140$

Alternative 2:—$(-300 + 275 + 310)/3 = 95$

Alternative 3:—$(-350 + 260 + 375)/3 = 95$

In the above case, using Laplace criterion alternative 1 is chosen.

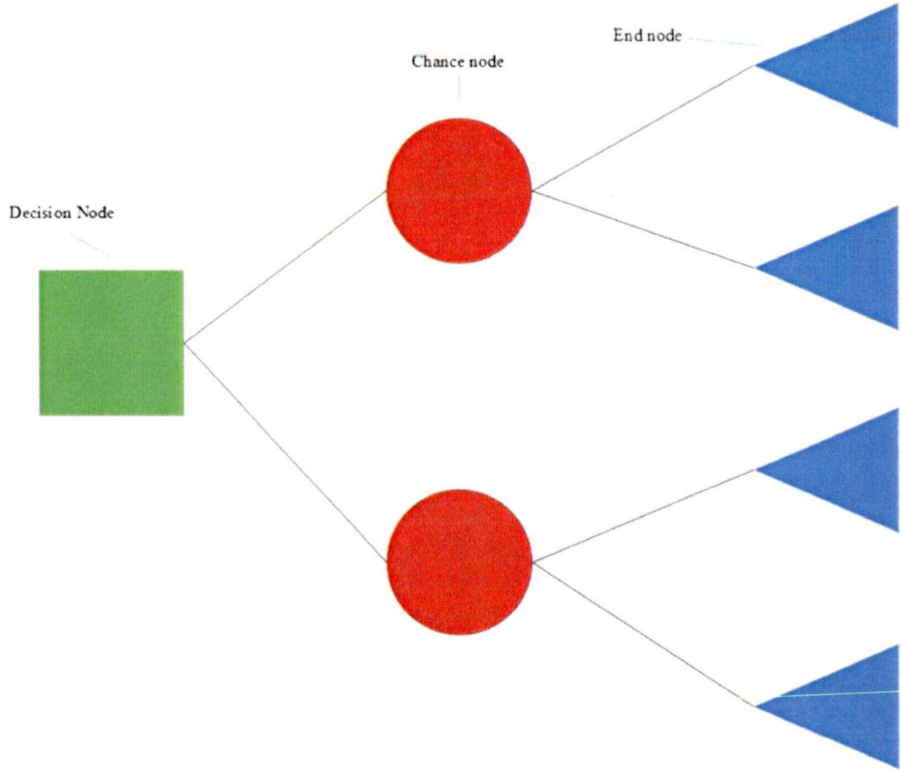

Fig. 3.4 Decision trees

In the *maximin criterion,* the least value for each alternative among the available iterations is chosen. In the next step, the alternative that has the largest value from among the previously chosen values is selected. In short, the alternative that has the best worst case is chosen for analysis. From the example given in Table 3.1, we get,

Alternative 1: −200
Alternative 2: −300
Alternative 3: −350

The alternative 1 would be selected as the outcome according to the maximin criterion as it has the best worst-case value.

In *maximax criterion,* the greatest value for each alternative among the available iterations is chosen. In the next step, the alternative that has the largest value from among the previously chosen values is selected. In short, the alternative that has the greatest best case is chosen for analysis. From the example given above, we get,

Alternative 1: 320
Alternative 2: 310
Alternative 3: 375

The alternative would be selected as the outcome according to the maximax criterion as it has the greatest best-case value.

In *Hurwicz criterion* [4], a compromise between maximin and maximax is established by striking a balance between the two. To achieve this coefficient, usually subjective is used to define the weighting for maximin or maximax. 'α,' called the coefficient of pessimism, is used to define the weighting on maximin and '$1-\alpha$' defines the weighting on maximax. The formula is given as:

$$\text{Hurwicz criterion} = \text{maximin} * \alpha + \text{maximax} * (1 - \alpha)$$

Also, in certain cases, when 'α' is called the coefficient of optimism then '1-α' defines the weighting on maximin and 'α' defines the weighting on maximax.

In *minimax regret*, a regret table is defined with the aim of reducing the decision's worst-case regret. The decision matrix and the corresponding regret table are shown in Tables 3.3 and 3.4, respectively. In the decision matrix, the best-case outcome for each alternative is given in the last row. The regret values are calculated by subtracting the individual outcome with the best-case outcome given in the last row. The worst regret for each alternative is shown in the last column. Therefore, the alternative 1 with the least regret value (55) should be chosen according to the minimax regret mechanism. A few rules for making decisions under risk are discussed below.

In *maximize expected value*, the option or alternative that has the highest expected value is chosen. The formula is given as follows.

$$\text{Expected value} = \sum (\text{Probability} * \text{outcome})$$

Maximize expected utility is a risk averse mechanism where unlike the maximize expected outcome, we simply target the increase in expected payoff without considering the risk factors. In such cases, we make of expected utility function that helps

Table 3.3 Decision Matrix with best outcomes

	Case 1	Case 2	Case 3
Alternative 1	−200	250	320
Alternative 2	−300	275	310
Alternative 3	−350	260	375
Best	−200	275	375

Table 3.4 Regret table

	Case 1	Case 2	Case 3	Worst regret
Alternative 1	0	25	55	55
Alternative 2	100	0	65	100
Alternative 3	150	15	0	150

to convert the payoffs to utility values. The shape of the utility curve is affected with the specification of the risk tolerance factor R. When $R = 0$, it indicates the risk neutral nature, $R > 0$ indicates risk averse and $R < 0$ indicates risk seeking.

In *most probable outcome*, the highest valued alternative from the case with the maximum probability is chosen for analysis. In the decision matrix shown in Table 3.1, the case 2 has the maximum probability with value 0.55 with alternatives as,

Alternative 1: 250
Alternative 2: 275
Alternative 3: 260

Therefore, we should choose alternative 2.

In *composite criteria*, we have a set of criteria that would be applied at each stage to eliminate alternatives and arrive at a final decision. For example, we could choose from a set of alternatives, the ones that have a maximum loss of only 100 and nothing more. This would result in a filtered set of alternatives upon which we can apply more criteria, or we could apply the known methods for analysis such as maximax, highest mean value and lowest variance alternatives, as the criteria for decision making.

Other basic yet important aspects of probability and statistics such as measures of central tendency, standard deviation, covariance, correlation, conditional probability, Bayes' theorem and the like play a major role in data analytics of which most of them have been discussed in Chapter 6.

3.1.4 Multicriteria Decision Analysis

Multicriteria decision analysis (MCDA) [5] is a branch of operational research used in decision making where multiple criteria are considered while making decisions. This holds good in real-world scenarios, wherein we have conflicting goals and constraints. It makes use of mathematical and other computational tools to arrive the best decision. MCDA includes two categories, namely the multiattribute decision making (MADM) and multiobjective decision making (MODM). Another categorization of MCDA includes the multiattribute utility theory (MAUT) and outranking methods. In MAUT methods, we calculate the utility of an alternative using a utility function. Outranking methods do a pairwise comparison of alternatives and identify the best alternative with the highest rank. The various MCDA methods are discussed below.

In *Analytic Hierarchical Process (AHP)*, complex problems are represented in a hierarchical fashion using pairwise comparison. An M x N matrix is used for analysis where M represents the alternatives and N represents the criteria.

The *Analytic Network process (ANP)* is an extension of AHP, wherein the problem is represented as a network rather than a hierarchy. This is done due to the fact that in many real-world problems there exists an element of interaction and dependence between the various levels that necessitate the inclusion of feedback. Here, the problems are modeled using ratio scale measurements based upon pairwise comparison.

The interdependence and communication are controlled in ANP using composite weights and super matrix.

In *Technique for Order of Preferences by Similarity to Ideal Solutions (TOPSIS),* accurate values are used to represent the criteria's performance ratings and weights. The methodology was further expanded to include fuzzy scenarios. In fuzzy TOPSIS, a fuzzy positive ideal solution (FPIS) and fuzzy negative ideal solution (FNIS) are defined, and then, the distance of every alternative from FPIS and FNIS is calculated. Finally, the alternatives are ranked in order based upon the closeness coefficient.

In *Elimination and Choice Expressing Reality (ELECTRE),* the alternatives under each criterion are compared pairwise to deal with the existing outranking relations. This method helps in achieving a smaller set of alternatives by eliminating the alternatives that are less effective and dominated by others. The ranks of alternatives are obtained with the help of graphs.

Fuzzy set theory aids in situations with unclear or insufficient data/information. Linguistic variables are used to describe fuzzy terms after which the mapping of linguistic values to numeric values is done.

3.2 Motivation

Decision making is an important and crucial activity in the field of engineering. A good product design requires a good decision-making model in situations involving risk and uncertainty. A good understanding of the decision-making process for the stakeholders is also required in the process of design and development of the product. Engineers typically differ from scientist as engineers are greatly involved in the process of decision making. Engineering decisions usually have important and high consequences to the society, organizations and the engineers themselves.

The effects of the decisions made by the engineers impact the environment and safety of the common man and the organization's performance and profitability [6]. Decisions shape our lives. Mathematical models explain the shift in information and establishes the balance among alternatives in any decision-making process. The overwhelming inflow of data is organized with the help of computer programs and algorithms that are built upon well-defined mathematical models. Under situations of uncertainty and risk, mathematical models help in bringing clarity and present the alternatives and their effects in an ordered fashion. Larger goals are made achievable with control over decisions in the lower level with the help of quantitative mathematical models. Expressing the decision process and model in a mathematical form helps the engineers in assessing the risks and outcomes before implementation and taking necessary actions Due to the massive influence that the engineer's decisions have in all aspects, it is important that the engineers have a good understanding of the mathematics involved in decision making, as a qualitative analysis will help in arriving at concrete and better outcomes.

3.3 Case Studies

Case study 1:

Cloud computing is one of the most cutting-edge technological advancements in today's IT business. The service-oriented architecture is responsible for cloud computing or simply cloud. Every company is attempting to take use of the cloud's benefits to decrease costs in infrastructure, network, hardware, software and other areas, but also to give end users with a seamless experience while maintaining scalability. Multitenancy is a concept that allows cloud service providers to save money by providing services to numerous users/companies at the same time using a shared resource. There are currently multiple cloud service providers on the market, all of which are rapidly altering and reorienting themselves in response to market need. To win market share, cloud service providers attempt to give the most up-to-date technologies to end users/customers while lowering costs. In such a situation, cloud consumers will find it incredibly difficult to choose the appropriate service provider for their needs. It is also growing more difficult to choose a deployment methodology among the many options available. The deployment models are appropriate for a variety of businesses. Varied deployment models have different criteria that are not tailored to an organization. As a cloud customer, deciding on a model and selecting the right service provider might be tricky, which has been addressed in [1], which is a computer engineering application. In the article, the authors have used the multicriteria decision-making approach to identify the most suitable service provider from the top four organizations and to select the deployment model that best meets the needs. The application of fuzzy TOPSIS [8] mechanism to evaluate the cloud service providers is as follows. In this paper, the three cloud service providers to be evaluated are given to be A, I and G. They are evaluated by three experts $D1$, $D2$ and $D3$. The linguistic variables used by the decision makers are mentioned in a seven-point scale in Table 3.5.

The criteria assessed by the decision makers $D1$, $D2$ and $D3$ are given in Table 3.6.

Considering the values of the linguistic variables, the fuzzy weights of each criteria j are given by

Table 3.5 Seven-point scale for linguistic variables

Very low (VL)	(0,0,0.1)
Low (P)	(0,0.1,0.3)
Medium low (ML)	(0.1,0.3,0.05)
Medium (M)	(0.3,0.5,0.7)
Medium high (MH)	(0.5,0.7,0.9)
High (H)	(0.7,0.9,1.0)
Very high (VH)	(0.9,1.0,1.0)

Table 3.6 Criteria assessed by decision makers

Feature name	*D*1	*D*2	*D*3
Business size support	*H*	VH	VH
Support for versatile industries	VH	*H*	*H*
Control interface features	*H*	*H*	*H*
Availability of support services	VH	VH	VH
Server operating system types	*H*	*H*	VH
Preconfigured operating systems	MH	MH	MH
Available runtimes	MH	*H*	MH
Middleware	*H*	MH	MH
Native databases	VH	VH	*H*

$$w_j = \frac{1}{3}\left[w_j^{(1)} + w_j^{(2)} + w_j^{(3)}\right]$$

Therefore,

$$\begin{aligned} w_1 &= \frac{1}{3}[G + VG + VG] \\ &= \frac{1}{3}[(0.7, 0.9, 1.0) + (0.9, 1.0, 1.0) + (0.9, 1.0, 1.0)] \\ &= \frac{1}{3}[2.5, 2.9, 3] \\ &= (0.83, 0.97, 1) \end{aligned}$$

Following the above method, we can find the values of $W2$, $W3$, ...$W9$ Table 3.7 shows the seven-point linguistic scale after evaluation of the three cloud service providers by the three decision makers.

A snippet of the cloud service provider features along with the analysis provided by the decision makers is shown in Table 3.8.

The evaluation for the cloud service provider for feature $F1$ is given by

Table 3.7 Seven-point linguistic scale for the cloud service providers

Very poor (VP)	(0,0,1)
Poor (*P*)	(0,1,3)
Medium poor (MP)	(1,3,5)
Fair (*F*)	(3,5,7)
Medium good (MG)	(5,7,9)
Good (*G*)	(7,9,10)
Very good (VG)	(9,10,10)

Table 3.8 Cloud service provider features and decision maker's analysis

Feature name	Cloud Providers	Decision makers		
		*D*1	*D*2	*D*3
Business size support (*F*1)	AWS	*G*	VG	VG
	IB	VG	*G*	*G*
	GCE	VG	VG	BG

$$
\begin{aligned}
x_{11} &= \frac{1}{3}[G + \mathrm{VG} + \mathrm{VG}] \\
&= \frac{1}{3}[(7, 9, 10) + (9, 10, 10) + (9, 10, 10)] \\
&= \frac{1}{3}[25, 29, 3] \\
&= (8.3, 9.6, 10)
\end{aligned}
$$

Similarly, the evaluation for all the cloud service providers and their features is done. The normalized fuzzy decision matrix $v = \left(v_{ij}\right)$ where $v_{ij} = \left(r_{ij}\right).\left(w_j\right)$. The fuzzy positive and fuzzy negative ideal solutions are given by

$$P^* = \left(V_1^*, V_2^*, \ldots V_9^*\right)$$

$N = \left(\overline{V_1}, \overline{V_2} \ldots \overline{V_9}\right)$
where, $V_j^* = (1, 1, 1)\overline{V_j} = (0, 0, 0)$.
The distance of the alternatives from the fuzzy positive is given by

$$d_i^+ = \sum_{j=1}^{n} d\left(V_{ij}, V_j^*\right)$$

The distance of the alternatives from (0,0,0) is also calculated. The calculated separation measures from the fuzzy positive and negative solutions are given in Table 3.9.

The closeness coefficient CC_i is used to calculate the ranking order of the cloud service providers and is given by $\frac{d_i^+}{(d_i^+ + d_i^-)}$. From the calculations, it is found that A ranks first followed by G and finally I.

Case Study 2:

The authors in [7] have demonstrated a risk analysis approach toward power systems planning. Three objective functions for various load scenarios have been

Table 3.9 Separation measures

Cloud providers	d_1^+	d_1^-
AWS	3.6759	6.0917

considered, namely costs incurred due to loss and investment, reliability and solution inadequacy. The decision criteria assist the decision makers in arriving at the most appropriate planning alternative. The decision criteria used in this case for evaluation are the minimum expected value criterion and the minimax weighted regret criterion. Using the minimum expected value criterion, he preferred solution represented by w^{opt} can be evaluated as follows,

$$w^{\text{opt}} = \text{argmin}_{i=1,\ldots,N_A}\left\{\sum_{k=1,\ldots,K} p_k b_{ik}\right\}$$

The minimax weighted regret criterion helps in minimizing the regret that the decision maker might consider. For a kth scenario with objective function b_k^{opt} the decision maker's regret felt for having chosen an alternative i, in the occurrence of the kth solution, is given by

$$r_{ik} = b_{ik} - b_k^{\text{opt}}$$

The preferred solution is obtained upon solving for

$$w_r^{\text{opt}} = \text{argmin}_{i=1,\ldots,N_A}\left\{\sum_{k=1,\ldots,K} p_k r_{ik}\right\}$$

3.4 Summary

Decision making is an inevitable aspect in all business areas. Engineers and managers encounter decision-making challenges in all industrial processes, scientific and technical streams. In this chapter, we have discussed the most common mathematical models for decision making that helps in enlightening the students in adopting and applying suitable mathematical concepts toward modeling the decision process. The concepts discussed in the chapter will help students develop skills of mathematical modeling for complex risk-associated problems, processes and strategies in organizations and society. Case studies presented in this chapter will enable the researchers and students understand the applicability of the above-discussed mathematical concepts to real-world applications.

References

1. Filieri A, Maggio M (2016) Technical briefing: control theory for software engineering. In: 2016 IEEE/ACM 38th international conference on software engineering companion (ICSE-C), pp 908–909
2. Shukla SK (2005) Teaching game theory for computer engineering. In: 2005 IEEE international conference on microelectronic systems education (MSE'05), pp 41–42. https://doi.org/10.1109/MSE.2005.54
3. Goerdin SAV, Smit JJ, Mehairjan RPY (2015) Monte Carlo simulation applied to support risk-based decision making in electricity distribution networks. IEEE Eindhoven Power Tech 2015:1–5. https://doi.org/10.1109/PTC.2015.7232494
4. Rizaev I, Takhavova E (2016) Statistical methods of making decision on repair of technical means. In: 2016 2nd international conference on industrial engineering, applications and manufacturing (ICIEAM), pp 1–4. https://doi.org/10.1109/ICIEAM.2016.7911647
5. Velasquez M, Hester PT (2013) An analysis of multi-criteria decision making methods. Int J Operat Res 10(2):56–66
6. Balters S, Steinert M (2014) Decision-making in engineering—a call for affective engineering dimensions in applied engineering design and design sciences. In: Proceedings of the 2014 international conference on innovative design and manufacturing (ICIDM, pp 11–15. https://doi.org/10.1109/IDAM.2014.6912663
7. Miranda V, Proenca LM (1998) Why risk analysis outperforms probabilistic choice as the effective decision support paradigm for power system planning. IEEE Trans Power Syst 13(2):643–648. https://doi.org/10.1109/59.667394
8. Basu A, Ghosh S (2018) Implementing fuzzy TOPSIS in cloud type and service provider selection. Adv. Fuzzy Syst 2018:1–12, Article ID 2503895. https://doi.org/10.1155/2018/2503895

Chapter 4
Optimization Problems

4.1 Introduction

Engineering of an every product involves various stages including concept, analysis, design, manufacturing, testing, marketing sales, etc. At every stage, the development product or process is sprouting lot of interaction between man, machine, money and resources. Aim of the entire system is to produce the maximum output with available resources with minimum time. During the entire process, it may happen that the repetitive processes can occur which eventually may delay in process of product development. It can be minimized by studying the process in detail and adopting solutions obtained from some mathematical techniques.

In the design of complex systems, large data is required. Data collection, segregation sorting it is a challenging part of the design. However, the way for such problems can be addressed using advance computing techniques. In the broadest sense, optimization is the process of making something as efficient as possible. As engineers, we must constantly design devices and products that perform tasks efficiently and at the lowest possible cost. As a result, engineers are constantly confronted with optimization problems that attempt to balance performance and constraints. Furthermore, scientists and engineers are interested in optimal phenomena of many things like minimizing the weight of aeroplane structure, maximizing of the profit of a manufacturing unit, detection maximum energy from neutrino, etc.

4.1.1 *Mathematical Modeling*

The process of creating a mathematical representation of a real-world scenario in order to make a prediction or provide insight is referred to as mathematical modeling. There is a distinction between using a formula and creating a mathematical relationship. The mathematical models generally express the essential feature of the physical system or process in mathematical term. These models are mainly governed by

P. N. Mahalle et al., *Foundations of Mathematical Modelling for Engineering Problem Solving*, Studies in Autonomic, Data-driven and Industrial Computing,
https://doi.org/10.1007/978-981-19-8828-8_4

the physical or chemical phenomenon of the system. To develop the mathematical model, one must understand the physics of the process or system, and the parameters (dependent or independent) involved. The general mathematical form of any system of process is described as follows:

$$\text{Dependent variable} = f(\text{forcingfunction, parameters, independentvariables})$$

Dependent variable is the output of the system represented by the all the elements involved in the process. Forcing function mainly associated with the physics of the system, process. The independent variables are typically dimensions, such as time and space, along which the behavior of the system is determined. The parameters reflect the properties or composition of the system. Examples are Newton's law of cooling, Ohm's law, Bernoulli's equation, steady-state head conduction, etc.

In many applications, creation of mathematical equation is not the ultimate solution. Further process is to find out its maximum or minimum value of the function. Here, the optimization comes in to picture. Optimization is the mathematical process of determining the maxima and minima of a function that is dependent on one or more variables. The goal is to find the variable values that produce maxima or minima for the function.

4.1.2 History Development of Optimization

Optimization has been studied since long decade to get remarkable and measurable output from the available recourses. Following flowchart shows the development in optimization techniques.

Engineering application of optimization

Optimization is one of the important parts of any engineer applications and some of them are as follows:

- Minimum cost design of civil structures such as foundations, frames, towers, bridges, dams and chimneys.
- Minimum weight design of aerospace structures.
- Optimal electrical machines such as motors, generators and transformers design.
- Optimal design of electrical networks.
- Optimal production planning, control and scheduling.
- Analyzing statistical data and building empirical models to obtain the most accurate representations of statistical phenomena.
- Optimal design of chemical processing equipment and plants.
- Designing optimal piping networks for process industries.
- Determining the optimal trajectory of the spacecraft.
- Designing minimum weight structures for seismic, wind and other types of random loads.

- Designing water resource systems to maximize profits.
- Optimal plastic design of construction.
- Selection of machining conditions in the machining process to minimize product costs.
- Design material handling equipment such as conveyors, trucks and cranes to minimize costs.
- Design pumps, turbines and heat transfer equipment for maximum efficiency.
- Designs to find the shortest routes needed to visit different cities in one of the travelers' tours.
- Optimal design of chemical processing equipment and plants.
- Designing optimal piping networks for process industries.
- Selection of industrial locations.
- Schedule maintenance and equipment replacement to reduce operating costs.
- Optimization of merchandise management processes.
- Allocating resources or services to multiple activities to maximize profits.
- Design pumps, turbines and heat transfer equipment for maximum efficiency.
- Optimal layout of linkages, cams, gears, machine tools and other machine parts.

4.1.3 Optimization

With the available conditions, making anything better is the process of optimization. In other terms, optimization describes the procedure of identifying the ideal answers to a certain issue. The following standard general form can be used to represent all optimal design problems:

Minimize or maximize the objective function $f(X)$

Subject to $g_i(X) \leq 0 \quad i = 1, 2, \ldots, p$ -inequality constraints
$h_j(X) = 0 \quad j = 1, 2, \ldots, q$ -equality constraints

$$X_{kL} \leq X_k \leq X_{kU} \quad k = 1, 2, \ldots, r$$

$f(X)$ is the objective function, $g_i(X)$ is inequality constraints, $h_j(X)$ is equality constraints. The vector X_k is the design vector whose value is to be determined to obtimize the objective function $f(X)$. X_{kL}andX_{kU} are the lower and upper bounds in which the optimum design of the function needs to be found out. It is also known as the side constraints. In the general case, the objective and constraint functions can be linear or nonlinear and can be explicit or implicit functions.

The majority of optimization algorithms take side constraints into account independently of equality and inequality constraints. By directly implementing the side constraints in the method, the side constraints can be managed effectively. Any of the side constraints will never be broken by a decent algorithm. Thus, an unconstrained optimization problem has no side constraints, but may contain equality or inequality requirements. One or more equality and/or inequality constraints, either with or without side constraints, are present in a constrained optimization problem. An inequality constraint can either be active or satisfied for a restricted optimization problem while an equality constraint can be either violated or satisfied. When $g_i(X) = 0$, an inequality constraint is activated.

Classification of Optimization Problems

Engineering problem of different nature is to be optimized using different optimization techniques. The classification of optimization problems is as follows:

- Classification based on objective function
  - Single-objective optimization
  - Multi-objective optimization
  - Pareto optimum

- Classification based on constraints
  - With or without constraints optimization
  - Equality or inequality constraints optimization

- Classification based on objective and constraints
  - Linear programming
  - Quadratic programming
  - Nonlinear programming
  - Geometric programming
  - Convex optimization
  - Non-smooth optimization

- Classification based on the nature of optimization
  - Global or local optimization
  - Deterministic or stochastic optimization

- Classification based on types of variable
  - Continuous variable optimization
  - Discrete variable optimization
  - Deterministic or stochastic optimization
  - Finite variable or function optimization
  - Evolutionary optimization
  - Genetic algorithm
  - Genetic programming
  - Differential evolution
  - Grey relational analysis
  - Evolutionary programming

4.2 Motivation

The usage of principles of nature to develop or construct objects useful to society is the underlying idea in the field of engineering. Irrespective of the field of engineering and the product developed, the engineer is always concerned about not just delivering a design that works at some nominal level, but in developing the best design that finds itself a place in the market. The process involved in controlling and confirming that the design is the best is called optimization. Optimization enables engineers to accomplish the finest designs under the given criteria and constraints. This helps in improving productivity, strength, reliability in the technology, the longevity of products, efficiency and utilization of resources. Optimization guides us on efficient resource usage in our journey to achieve the goal while crossing many hurdles and conditions. The objective of the optimization model in a production plant would be to reduce the cost incurred using the resources such as labor and raw materials. The expertise in establishing a balance while dealing with constraints and criteria in the process of selecting an appropriate solution is an important skill that is vital for all engineers irrespective of their field of operation. This will help in arriving at better designs and solutions and reduce the overhead of costs incurred and re-arrive at better designs and solutions and reduce overhead source utilization.

Generally, the optimization process is implicit, which includes various aspects such as judgment based on experiences, opinions of others and modeling. Using such factors helps the engineer to make better design decisions leading to an optimal design. However, in situations that include many variables to be evaluated and many conflicting constraints and objectives, the experience-based or qualitative optimization process may not lead to successful optimal designs. The complex interactions and a huge number of variables make the process difficult to be tracked. Hence, it is important to quantitatively carry out the optimization procedure by adopting the mechanisms discussed in the chapter.

4.3 Case Study

In this section, the machining parameter optimization using grey relational analysis techniques is discussed in detail. Turning process is used for machining. The work material is considered as SS 304 (austenitic grade stainless steel). The cutting tool details are given in Table 4.1 and cutting parameters details are given in Table 4.2.

Table 4.1 Coating tool details

Coating materials	Coating technique	Tool
AlTiCrN	CAE	T1
AlTiN	DCMS	T2
AlTiCrN	HPPMS	T3

Table 4.2 Test conditions used for evaluation of machining performance

Expt. No	Input factors		
	Cutting speed (m/min)	Feed (mm/rev)	Depth of cut (mm)
1	140	0.08	1
2	140	0.14	1
3	140	0.2	1
4	140	0.26	1
5	200	0.08	1
6	200	0.14	1
7	200	0.2	1
8	200	0.26	1
9	260	0.08	1
10	260	0.14	1
11	260	0.2	1
12	260	0.26	1
13	320	0.08	1
14	320	0.14	1
15	320	0.2	1
16	320	0.26	1

Grey Relational Analysis

The theory of grey analysis was initially developed and presented by Deng in 1982. According to grey system theory [2], an actual system is considered ‘grey,’ meaning that some information can be known while some may not be known precisely or at all. This theory converts a multiple response optimization problem into a single response optimization situation with the objective function of overall grey relational grade. The grey system theory has been successfully used in various fields, including industry, agriculture, economics, mechanical and civil engineering. This method can effectively solve problems that are uncertain or incomplete or which involve systems with incomplete information, using system relational analysis, model construction and forecasting and decision analysis. From an engineering point of view, minimizing manufacturing resources is the main concern during design and production [2]. The smaller the total manufacturing cost is, the better the optimization result is. The objective of optimization is to make the difference between the objective sequence and the standard sequence as small as possible, that is, the smaller the grey difference degree, the better the results. Although the system to be studied is complex in appearance and only discrete data and incomplete information can be obtained from it, its factors are related to each other and there must exist some inner hidden rules. As its two extremes, a white system means that all of its information is known and a black system means that no information is known. Grey theory presents a degree of grey correlation to describe the amount of similarity in the developing trends of different things or different factors. The greater the degree of grey correlation, the more similar the things are, and vice versa [2].

In turning process, the machining performance mainly depends on the process, workpiece and tool-related parameters like cutting speed, feed, depth of cut and edge geometry of the cutting tool and cutting environment, etc. These parameters have significant effect on the performance measures such as cutting forces, surface roughness and tool wear. If there are multiple response variables, for the same conditions of independent variables, the optimum parameter cannot found by Taguchi method because Taguchi method provides separate optimum parametric conditions for each response variable. However, grey relational analysis ranks the experiments based on the increasing order of their grey relational grade (GRG). This GRG can be used to identify the most influencing factors affecting the response variables. In the present study, the method is used to determine optimum parameter for single layer coated tool.

Methodology of Grey Analysis

The procedure for obtaining the solution of Taguchi-GRA optimization is given as follows:

Step 1 Identify input parameters that influence the multiple output variables.
Step 2 Selection of Taguchi design matrix and conducting the experiments.
Step 3 Selection of quality characteristics for each output variable.
Step 4 Normalize all response variables (grey relational generating).
Step 5 Determine the values of comparability sequences.
Step 6 Calculate the grey relational coefficient (GRC) for each output.
Step 7 Calculate the grey relational grade by mean value of GRCs.
Step 8 Recognize the noticeable and unnoticeable variable factors and select.
Step 9 Prediction of grey relational grade for optimal machining parameters.

The first three steps were performing as per the Taguchi analysis, and the experimental results were considered for the grey relational analysis. The subsequent steps used in GRA are described in detail as follows:

Step 4: Normalization of response variables

The normalization of the response variables was performed to prepare raw data for analysis where the original sequence is transferred to a comparable sequence. Linear normalization is usually required since the range and unit in one data sequence may differ from the others. The process of linear normalization of the original sequence is a problem of 'smaller-the-better' type [3]. Hence, 'the smaller-the-better' normalization formula was used to transfer original sequence to comparable sequence and is given below.

$$x_i^*(k) = \frac{\max x_i^{(o)}(k) - x_i^{(o)}(k)}{\max x_i^{(o)}(k) - \min x_i^{(o)}(k)} \tag{4.1}$$

If the target value of the original sequence is infinite, then the larger-the-better quality characteristic is used to normalize the original sequence using the following equation [3]

$$x_i^*(k) = \frac{x_i^{(o)}(k) - \min x_i^{(o)}(k)}{\max x_i^{(o)}(k) - \min x_i^{(o)}(k)} \tag{4.2}$$

Step 5: Determination of Deviation Sequences, $\Delta 0_i(k)$

The deviation sequence, $\Delta 0_i(k)$, is the absolute difference between the reference sequence $x_0{}^*(k)$ and the comparability sequence $x_i{}^*(k)$ after normalization. The value of $x_0{}^*(k)$ was considered as 1. It is determined using Eq. 'A3' as given below [3].

$$\Delta 0_i(k) = \left|x_0^*(k) - x_i^*(k)\right| \tag{4.3}$$

Step 6: Calculation of Grey Relational Coefficient, GRC

Grey relational coefficients (GRC) for all the sequences express the relationship between the ideal (best) and actual normalized response variables [4]. If the two sequences agree at all points, then their grey relational coefficient is one. The grey relational coefficient $\gamma(x_0(k), x_i(k))$ can be expressed by Eq. 'A4.'

$$\gamma(x_0(k), x_i(k)) = \frac{\Delta \min + \zeta \Delta \max}{\Delta_{0i}(k) + \zeta \Delta \max} \tag{4.4}$$

where $\Delta \min$ is the smallest value of $\Delta 0_i(k) = \min_i \min_k |x_0*(k) - x_i*(k)|$ and $\Delta \max$ is the largest value of $\Delta 0_i(k) = \max_i \max_k |x_0*(k) - x_i*(k)|$, $x_0{}^*(k)$ is the ideal normalized S/N ratio, $x_i{}^*(k)$ is the normalized comparability sequence and ζ is the distinguishing coefficient. The value of (ζ) is 0.5 for all response variables and is substituted in Eq. 'A4.' The GRC for all the experimental runs is calculated.

Step 7: Determination of grey relational grade, GRG

The overall evaluation of the multiple performance characteristics is based on the grey relational grade. The grey relational grade is an average sum of the grey relational coefficient, which is defined as follows:

$$\gamma(x_0, x_i) = \frac{1}{m}\sum_{i=1}^{m} \gamma(x_0(k), x_i(k)) \tag{4.5}$$

where $\gamma(x_0, x_i)$ the grey relational grade for the jth experiment and m is the number of performance characteristics.

The grey relational grade $\gamma(x_0, x_i)$ represents the level of correlation between the reference sequence and the comparability sequence. If the two sequences agree at all points, then their grey relational coefficient is 1 everywhere, and therefore, their grey relational grade is equal to 1. The grey relational grade was determined

Table 4.3 Grey relational coefficient (GRC) values for cutting force

Expt. No	Tool	Cutting force (N)	Xi*(k)	0i	GRC
1	*T*1	396	0.82	0.18	0.740
2	*T*1	588	0.44	0.56	0.472
3	*T*1	694	0.23	0.77	0.393
4	*T*1	808	0.00	1.00	0.333
5	*T*1	368	0.88	0.12	0.806
6	*T*1	536	0.54	0.46	0.523
7	*T*1	642	0.33	0.67	0.428
8	*T*1	780	0.06	0.94	0.346
9	*T*1	354	0.91	0.09	0.845
10	*T*1	524	0.57	0.43	0.536
11	*T*1	607	0.40	0.60	0.455
12	*T*1	754	0.11	0.89	0.359
13	*T*1	350	0.92	0.08	0.856
14	*T*1	510	0.60	0.40	0.553
15	*T*1	579	0.46	0.54	0.480
16	*T*1	768	0.08	0.92	0.352
17	*T*2	384	0.85	0.15	0.767
18	*T*2	556	0.50	0.50	0.502
19	T2	652	0.31	0.69	0.421
20	*T*2	772	0.07	0.93	0.350
21	*T*2	360	0.90	0.10	0.828
22	*T*2	518	0.58	0.42	0.543
23	*T*2	631	0.35	0.65	0.436
24	*T*2	754	0.11	0.89	0.359
25	*T*2	346	0.92	0.08	0.868
26	*T*2	512	0.59	0.41	0.551
27	*T*2	601	0.41	0.59	0.460
28	*T*2	718	0.18	0.82	0.379
29	*T*2	334	0.95	0.05	0.906
30	*T*2	498	0.62	0.38	0.568
31	*T*2	573	0.47	0.53	0.485
32	*T*2	714	0.19	0.81	0.381
33	*T*3	372	0.87	0.13	0.796
34	*T*3	508	0.60	0.40	0.556
35	*T*3	624	0.37	0.63	0.442
36	*T*3	712	0.19	0.81	0.382
37	*T*3	342	0.93	0.07	0.880
38	*T*3	452	0.71	0.29	0.635

(continued)

Table 4.3 (continued)

Expt. No	Tool	Cutting force (N)	Xi*(k)	0i	GRC
39	*T*3	570	0.48	0.52	0.488
40	*T*3	650	0.32	0.68	0.422
41	*T*3	314	0.99	0.01	0.977
42	*T*3	428	0.76	0.24	0.676
43	*T*3	524	0.57	0.43	0.536
44	*T*3	618	0.38	0.62	0.446
45	*T*3	308	1.00	0.00	1.000
46	*T*3	398	0.82	0.18	0.735
47	*T*3	490	0.64	0.36	0.579
48	*T*3	576	0.46	0.54	0.483

Table 4.4 Grey relational coefficient (GRC) values for surface roughness

Expt. No	Tool	Surface roughness (μm)	$Xi^*(k)$	$0i$	GRC
1	*T*1	2.12	0.67	0.33	0.604
2	*T*1	2.71	0.49	0.51	0.494
3	*T*1	3.52	0.23	0.77	0.395
4	*T*1	4.27	0.00	1.00	0.333
5	*T*1	1.82	0.77	0.23	0.681
6	*T*1	2.46	0.57	0.43	0.535
7	*T*1	3.08	0.37	0.63	0.443
8	*T*1	3.7	0.18	0.82	0.378
9	*T*1	1.61	0.83	0.17	0.748
10	*T*1	2.18	0.65	0.35	0.590
11	*T*1	2.68	0.50	0.50	0.498
12	*T*1	3.41	0.27	0.73	0.406
13	*T*1	1.58	0.84	0.16	0.758
14	*T*1	2.12	0.67	0.33	0.604
15	*T*1	2.62	0.52	0.48	0.508
16	*T*1	3.64	0.20	0.80	0.384
17	*T*2	1.89	0.74	0.26	0.661
18	*T*2	2.56	0.53	0.47	0.518
19	*T*2	3.04	0.38	0.62	0.448
20	*T*2	3.68	0.18	0.82	0.380
21	*T*2	1.63	0.83	0.18	0.741
22	*T*2	2.18	0.65	0.35	0.590
23	*T*2	2.73	0.48	0.52	0.491

(continued)

Table 4.4 (continued)

Expt. No	Tool	Surface roughness (μm)	$Xi^*(k)$	$0i$	GRC
24	$T2$	3.21	0.33	0.67	0.428
25	$T2$	1.57	0.84	0.16	0.762
26	$T2$	2.03	0.70	0.30	0.625
27	$T2$	2.53	0.54	0.46	0.523
28	$T2$	2.98	0.40	0.60	0.456
29	$T2$	1.45	0.88	0.12	0.808
30	$T2$	1.98	0.72	0.28	0.637
31	$T2$	2.43	0.58	0.43	0.541
32	$T2$	3.12	0.36	0.64	0.438
33	$T3$	1.56	0.85	0.15	0.766
34	$T3$	2.15	0.66	0.34	0.597
35	$T3$	2.77	0.47	0.53	0.485
36	$T3$	3.38	0.28	0.72	0.409
37	$T3$	1.36	0.91	0.09	0.847
38	$T3$	1.88	0.75	0.25	0.664
39	$T3$	2.46	0.57	0.43	0.535
40	$T3$	2.96	0.41	0.59	0.458
41	$T3$	1.18	0.97	0.03	0.936
42	$T3$	1.7	0.80	0.20	0.717
43	$T3$	2.32	0.61	0.39	0.561
44	$T3$	2.74	0.48	0.52	0.489
45	$T3$	1.07	1.00	0.00	1.000
46	$T3$	1.58	0.84	0.16	0.758
47	$T3$	2.22	0.64	0.36	0.582
48	$T3$	2.59	0.53	0.48	0.513

by Eq. 'A5' and is shown in Table 5.6. A higher grey relational grade in Table 5.6 indicates that the corresponding condition is optimum. It is observed that experiment 3 has the highest grey relational grade.

Step 8: Determination of optimum parameters

The grey relational grade calculated for each sequence is taken as response for the further analysis. The larger-the-better quality characteristic was used for analyzing the GRG since larger value indicates the better performance of the process.

Step 9: Prediction of grey relational grade under optimum parameters setting

After evaluating the optimal parameter settings, the next step is to predict and verify the improvement of quality characteristics using the optimal parametric combination. The estimated grey relational grade $\hat{\gamma}$ using the optimal level of the machining parameters is calculated.

Table 4.5 Grey relational coefficient (GRC) values for tool wear

Expt. No	Tool	Tool wear (μm)	$Xi^*(k)$	0i	GRC
1	$T1$	33	0.88	0.13	0.800
2	$T1$	40	0.81	0.19	0.727
3	$T1$	52	0.71	0.29	0.629
4	$T1$	59	0.64	0.36	0.583
5	$T1$	50	0.72	0.28	0.644
6	$T1$	58	0.65	0.35	0.589
7	$T1$	72	0.53	0.47	0.514
8	$T1$	83	0.43	0.57	0.467
9	$T1$	63	0.61	0.39	0.560
10	$T1$	87	0.39	0.61	0.452
11	$T1$	98	0.29	0.71	0.415
12	$T1$	112	0.17	0.83	0.376
13	$T1$	85	0.41	0.59	0.459
14	$T1$	106	0.22	0.78	0.392
15	$T1$	118	0.12	0.88	0.361
16	$T1$	131	0.00	1.00	0.333
17	$T2$	29	0.91	0.09	0.848
18	$T2$	35	0.86	0.14	0.778
19	$T2$	40	0.81	0.19	0.727
20	$T2$	43	0.79	0.21	0.700
21	$T2$	43	0.79	0.21	0.700
22	$T2$	51	0.71	0.29	0.636
23	$T2$	58	0.65	0.35	0.589
24	$T2$	64	0.60	0.40	0.554
25	$T2$	57	0.66	0.34	0.596
26	$T2$	68	0.56	0.44	0.533
27	$T2$	76	0.49	0.51	0.496
28	$T2$	84	0.42	0.58	0.463
29	$T2$	73	0.52	0.48	0.509
30	$T2$	83	0.43	0.57	0.467
31	$T2$	91	0.36	0.64	0.438
32	$T2$	97	0.30	0.70	0.418
33	$T3$	19	1.00	0.00	1.000
34	$T3$	23	0.96	0.04	0.933
35	$T3$	26	0.94	0.06	0.889
36	$T3$	28	0.92	0.08	0.862
37	$T3$	29	0.91	0.09	0.848
38	$T3$	32	0.88	0.12	0.812

(continued)

Table 4.5 (continued)

Expt. No	Tool	Tool wear (μm)	$Xi^*(k)$	0i	GRC
39	$T3$	38	0.83	0.17	0.747
40	$T3$	40	0.81	0.19	0.727
41	$T3$	39	0.82	0.18	0.737
42	$T3$	44	0.78	0.22	0.691
43	$T3$	48	0.74	0.26	0.659
44	$T3$	51	0.71	0.29	0.636
45	$T3$	49	0.73	0.27	0.651
46	$T3$	55	0.68	0.32	0.609
47	$T3$	60	0.63	0.37	0.577
48	$T3$	64	0.60	0.40	0.554

Grey relational coefficient values for cutting force, surface roughness and average flank wear are calculated and it is given in Table A1, A2 and A3.

Calculation of Grey relational coefficient (GRC)

In the present study, linear normalization of the experimental results for cutting force, surface roughness and tool wear has done based on the smaller-the-better approach.

The experiments were conducted for different machining parameters as per Table 4.2 using each coated tool ($T1$, T2 and T3). Total 48 experiments were conducted out of which 1–16 experiments using $T1$, 17–32 using $T2$ and remaining using $T3$. Using Eqs. 4.1, 4.3 and 4.4, normalization of response ($xi^*(k)$), deviation sequence ($\Delta 0i$) and grey relational coefficient (GRC) for cutting force, surface roughness and tool wear were determined [6]. The $xi^*(k)$, Δ0i and GRC values for cutting force, surface roughness and tool wear given in Tables 4.3, 4.4 and 4.5.

Estimation of Grey relational Grade (GRG)

The overall evaluation of the multiple performance characteristics is based on the grey relational grade (GRG). GRG is an average sum of the grey relational coefficient [5]. A higher GRG indicates that the corresponding experimental condition is optimum [7]. The GRG values for all the forty-eight experiments are shown in Table 4.6.

The optimum parameters for $T1$, $T2$ and $T3$ tools are also obtained from GRG Table 4.6, and it is summarized in Table 4.7

The most significant parameter affecting performance characteristics is determined by comparing response value from GRG table. In the present study, three levels for tool coating, four levels for cutting speed and four levels for feed are considered [8]. The mean of GRG values for all these factors is obtained from mean effect plot (Fig. 4.1). These values are given in Table 4.8 level wise. The highest difference between maximum level and minimum level is considered the most significant parameter. It is clearly seen from Table 4.8 that feed (0.32295) has the strongest influence followed by tool coating (0.14883) on the machinability characteristics of SS 304 steel.

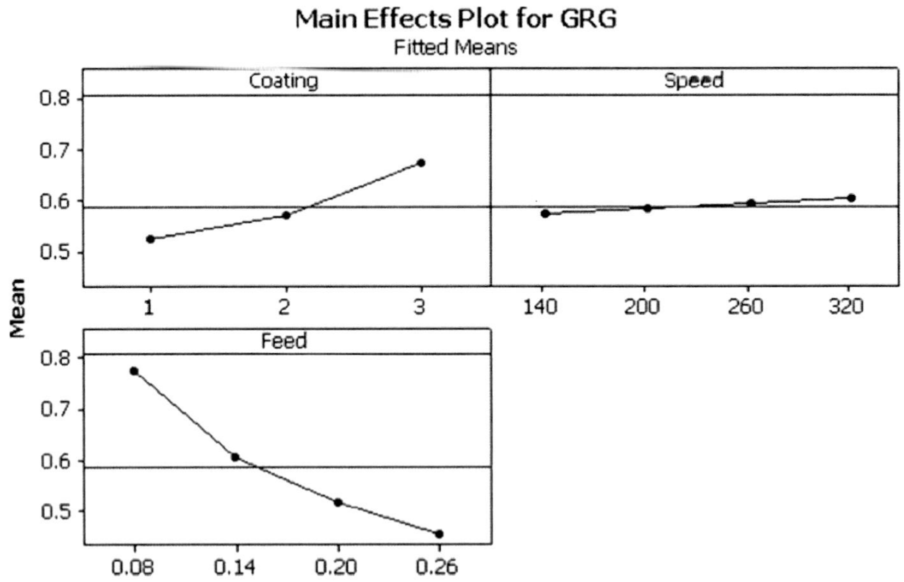

Fig. 4.1 Mean effect plot for grey Relational Grade

Analysis of Variance of Grey Relational Grade (GRG)

The ANOVA is performed to determine the factor, which significantly affects the multi-response characteristics in turning of SS 304 steel in different cutting condition using T1, T2 and T3 tools. The results of ANOVA for grey relational grade values

Table 4.6 Grey Relational Grades

Expt. No	Tool	GRC (Fc)	GRC (Ra)	GRC (V_B)	GRG
1	*T*1	0.740	0.604	0.800	0.714
2	*T*1	0.472	0.494	0.727	0.564
3	*T*1	0.393	0.395	0.629	0.472
4	*T*1	0.333	0.333	0.583	0.417
5	*T*1	0.806	0.681	0.644	0.710
6	*T*1	0.523	0.535	0.589	0.549
7	*T*1	0.428	0.443	0.514	0.462
8	*T*1	0.346	0.378	0.467	0.397
9	*T*1	0.845	0.748	0.560	0.717
10	*T*1	0.536	0.590	0.452	0.526
11	*T*1	0.455	0.498	0.415	0.456
12	*T*1	0.359	0.406	0.376	0.380
13	*T*1	0.856	0.758	0.459	0.691

(continued)

Table 4.6 (continued)

Expt. No	Tool	GRC (Fc)	GRC (Ra)	GRC (V_B)	GRG
14	*T*1	0.553	0.604	0.392	0.516
15	*T*1	0.480	0.508	0.361	0.450
16	*T*1	0.352	0.384	0.333	0.356
17	*T*2	0.767	0.661	0.848	0.759
18	*T*2	0.502	0.518	0.778	0.599
19	*T*2	0.421	0.448	0.727	0.532
20	*T*2	0.350	0.380	0.700	0.477
21	*T*2	0.828	0.741	0.700	0.756
22	*T*2	0.543	0.590	0.636	0.590
23	*T*2	0.436	0.491	0.589	0.506
24	*T*2	0.359	0.428	0.554	0.447
25	*T*2	0.868	0.762	0.596	0.742
26	*T*2	0.551	0.625	0.533	0.570
27	*T*2	0.460	0.523	0.496	0.493
28	*T*2	0.379	0.456	0.463	0.432
29	*T*2	0.906	0.808	0.509	0.741
30	*T*2	0.568	0.637	0.467	0.557
31	*T*2	0.485	0.541	0.438	0.488
32	*T*2	0.381	0.438	0.418	0.412
33	*T*3	0.796	0.766	1.000	0.854
34	*T*3	0.556	0.597	0.933	0.695
35	*T*3	0.442	0.485	0.889	0.605
36	*T*3	0.382	0.409	0.862	0.551
37	*T*3	0.880	0.847	0.848	0.858
38	*T*3	0.635	0.664	0.812	0.703
39	*T*3	0.488	0.535	0.747	0.590
40	*T*3	0.422	0.458	0.727	0.536
41	*T*3	0.977	0.936	0.737	0.883
42	*T*3	0.676	0.717	0.691	0.695
43	*T*3	0.536	0.561	0.659	0.586
44	*T*3	0.446	0.489	0.636	0.524
45	***T* 3**	1.000	1.000	0.651	0.884
46	*T*3	0.735	0.758	0.609	0.701
47	*T*3	0.579	0.582	0.577	0.579
48	*T*3	0.483	0.513	0.554	0.517

It is clearly observed from Table 4.6 that experiment number 45 has the highest grey relational grade. Therefore, experiment number 45, i.e., cutting speed 320 m/min, feed 0.08 mm/rev, depth of cut 1 mm and tool *T*3 (AlTiCrN-HPPMS), is the optimal process parameter for minimum cutting force, surface roughness and tool wear during turning of SS 304 steel (i.e., the best multi-performance characteristics) among the 48 experiments. From the GRG Table 4.6, it is further observed that *T*3 shows better performance compared to *T*2 and *T*1.

Table 4.7 Tool-wise optimum machining parameters

S. No.	Expt. No	Optimum process parameter			Tool
		Cutting speed (m/min)	Feed (mm/rev)	Depth of cut (mm)	
1	5	200	0.08	1	T1
2	25	260	0.08	1	T2
3	45	320	0.08	1	T3

Table 4.8 Mean of GRG values for different factors

Parameters	Average grey relational grade by factor level				(Max–Min)
	Level 1	Level 2	Level 3	Level 4	
Cutting Speed	0.5744	0.5837	0.59219	0.6033	0.0289
Feed	0.7759	0.6055	0.5182	0.4529	0.3230
Tool coating	0.5237	0.5688	0.6726	–	0.1489

Table 4.9 ANOVA for grey relational grade (GRG)

Source of variance	Degrees of freedom (DOF)	Sum of squares (SS)	Adjusted (SS)	Adjusted (MS)	*F* value	*P* value	% Contribution
Tool (Coating)	2	0.186	0.186	0.093	476.42	0.000	20.69
Speed	3	0.005	0.006	0.002	9.31	0.003	0.61
Feed	3	0.701	0.701	0.234	1195.33	0.000	77.86
Error	39	0.008	0.008	0.000			0.85
Total	47	0.901					100
$S = 0.0139856$ R-Sq $= 98.15\%$ R-Sq (adj) $= 97.98\%$							

with contribution percentage and F value and their interactions are shown in Table 4.9. The mean effect plot for GRG is shown in Fig. 4.1.

It is observed from the ANOVA that the feed and tool (coating) are statistically significant at 95% confidence interval effect on the variability of the performance characteristics of turning SS 304 steel. In addition, from the mean effect plot, it is observed that tool T3, cutting speed 320 m/min and feed 0.08 mm/rev are the optimum cutting conditions for SS 304 steel turning. This result is in line with grey relational analysis result. It is also observed that feed is the most influencing parameter followed by tool coating.

4.4 Summary

Optimization is an inevitable process in all engineering problems and in real life. Mathematical application to optimization process in product design and development will help in more accurate and measurable result-driven designs. The concepts discussed in the chapter will help students develop skills in mathematical modeling for complex problems, processes and strategies in organizations and society. Case studies presented in this chapter will enable the researchers and students to understand the applicability of the above-discussed mathematical concepts to real-world applications.

References

1. Ross SM (2014) Introduction to probability models, 11th edition, Elsevier Science
2. Chen Y, Huang M, Yao J et al (2003) Optimal concurrent tolerance based on the grey optimal approach. Int J Adv Manuf Technol 22:112–117. https://doi.org/10.1007/s00170-002-1448-2
3. Misal ND, Sadaiah M (2018) Multi-objective optimization of photochemical machining of inconel 601 using grey relational analysis. Mater Today Proc 5(2), Part 1:5591–5600, ISSN 2214-7853, https://doi.org/10.1016/j.matpr.2017.12.150
4. Bhosale SB, Pawade RS (2013) Grey relational parameter optimization in ultrasonic machining of ceramic composite (Al2O3/Zro2). J Association Eng India 83:63. https://doi.org/10.22485/jaei/2013/v83/i2/119906
5. Bhosale SB, Bhowmik S, Ray A (2021) Experimental analysis and parametric optimization of drilling process for ferrous clay composite using GRA-PCA approach. J Mater Res Technol 10:376–389, ISSN 2238–7854. https://doi.org/10.1016/j.jmrt.2020.12.032
6. Arun Premnath A (2015) Studies on machining parameters while milling particle reinforced hybrid (Al6061/Al_2O_3/Gr) MMC. Part Sci Technol 33(6):682–692. https://doi.org/10.1080/02726351.2015.1025457
7. Huang SJ, Chiu NH, Chen LW (2008) Integration of the grey relational analysis with genetic algorithm for software effort estimation, Eur. J. Operat. Res. 188(3):898–909, ISSN 0377-2217. https://doi.org/10.1016/j.ejor.2007.07.002
8. Suresh R, Basavarajappa S, Gaitonde VN, Samuel GL (2012) Machinability investigations on hardened AISI 4340 steel using coated carbide insert. Int. J. Refractory Metals Hard Mater 33:75–86, ISSN 0263-4368, https://doi.org/10.1016/j.ijrmhm.2012.02.019

Chapter 5
Delay Problems

5.1 Introduction

While we are pressing on toward improved and increased system performance expectations, delay in systems is also an important aspect that cannot be overlooked while modeling the system. In the field of engineering, time delays are inevitable. Indeed, many industrial processes incorporate aftereffect events into their internal dynamics. Also, sensors, actuators and field networks used in feedback loops repeatedly cause unavoidable delays. Studies focused on delay-related applications and systems are important which in turn will provide a broader avenue for researchers to explore. Additionally, control scientists and engineers seem to have brought time-delayed systems under the umbrella for research and development in the last few decades. One of the primary reasons is because the time-delayed systems in the field of engineering have often degraded the system performance and have led to serious instabilities in the original system. This has in turn made the activities such as controller design, system analysis and state estimation a difficult task. Therefore, it is necessary to address these drawbacks in delay systems by providing additional compensation and consideration to the delays in the systems.

Modeling a system to incorporate the delays incurred is a crucial aspect. This chapter aims to introduce the various mathematical aspects that support and can be used in modeling/representing the delay in systems. In this chapter, we will discuss majorly the usage of queuing theory and the delay differential equations to represent delays. Various methods for solving the delay differential equations are also mentioned along with appropriate case studies and examples.

P. N. Mahalle et al., *Foundations of Mathematical Modelling for Engineering Problem Solving*, Studies in Autonomic, Data-driven and Industrial Computing,
https://doi.org/10.1007/978-981-19-8828-8_5

5.1.1 Queuing Theory

Queuing theory is a mathematical discipline that is used to analyze and model the system of waiting and the associated queues that find its application in our every day-to-day life. In this section, we provide an overview and explain the various aspects related to queues supported by examples that show their application and usage. At the end of this section, readers will have sufficient information to be able to model a queuing application into one of the categories that have been presented. Also, derivation of the basic concepts of queues such as average waiting time, service time and so on should be an easy task for the readers after they have gained the fundamental concepts from this section. Before we start with the queuing theory, we must first understand the probability theory which is inevitable when considering the queuing system. Hence, we will look at the exponential and Poisson probability distributions [1, 2].

Exponential Distribution

It is impossible to forecast how much service time a customer will take in advance. As a result, a customer's service time is supposed to follow some probability distribution. In queuing models, the exponential distribution is the most employed. We frequently assume that a customer's service time is independent of all other customers' service times and follows an exponential distribution. Furthermore, as we will see in the following subsection, the duration between two consecutive arrivals to a queuing system is typically considered to follow an exponential distribution [3].

Let S denote (the random variable characterizing) an arbitrary customer's service time. Then S is said to have an exponential distribution if and only if for every $t \geq 0$:

$$\Pr\{S \leq t\} = 1 - e^{\mu t}$$

The parameter $\boldsymbol{\mu}$ is known as the **service rate**, and it represents the average number of clients that a single server can handle in the long term if the server never runs out of customers to serve . Figure 5.2 depicts this distribution for a variety of values such as m = 0.5, 1.0, 2.0

The following are some helpful information regarding the exponential distribution:

- mean service time $= E[S] = 1/\mu$,
- standard deviation of $S = \sigma\ [S] = 1/\mu$,
- coefficient of variation of $S = \mathrm{cv}[S] = \sigma[\mathrm{S}]/\mathrm{E}[S] = 1$.

Furthermore, the exponential distribution is said to be memoryless. This indicates that the remaining service time of a client who is presently in service follows the same exponential distribution as the service time of another customer who begins service now. Therefore, for all t, $s \geq 0$ we have

$$\Pr\{S \leq t | sS > s\} = 1 - e^{\mu t}$$

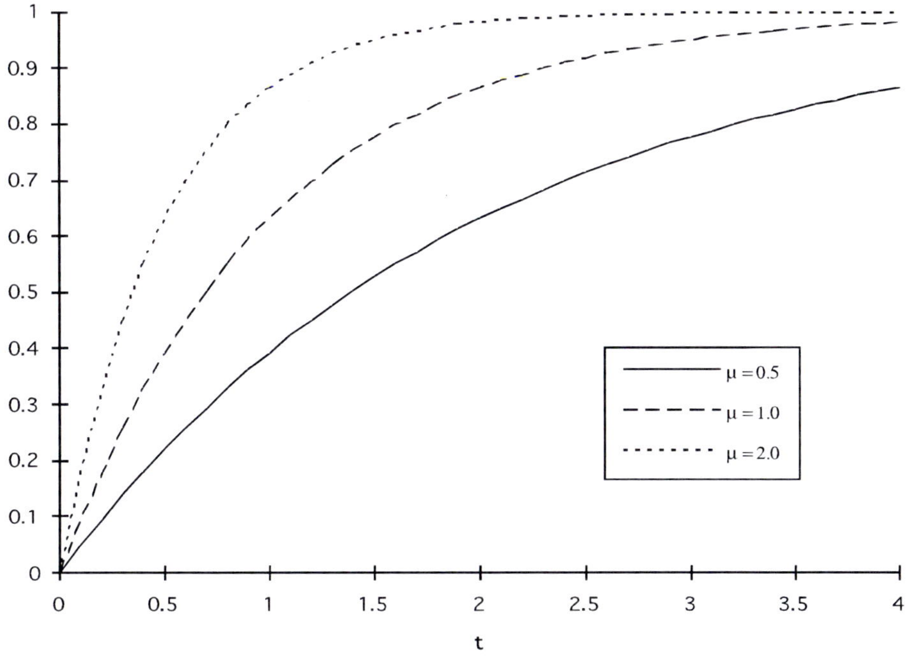

Fig. 5.1 Exponential cumulative distribution functions for $m = 0.5, 1.0, 2.0$ adopted from [1]

It is worth noting that the right-hand side of this equation is completely independent of s. This fact is what makes the exponential distribution useful in queuing models: it is not essential to keep track of how long a client has previously been in service because the remaining service time of the client will always follow the same distribution. The exponential distribution is represented by an '*M*' for memoryless in the generally used shorthand notation for queuing models. Figure 5.1 shows the exponential distribution for varying values of $m = 0.5, 1.0, 2.0$.

The Poisson Process

Arrivals at a service system are typically random and unpredictable. Even when a decent forecast of total arrivals is provided, there is frequently considerable uncertainty about the actual timing of arrivals. Consider the checkout line at a major hotel. Management has quite a good estimate of the number of guests who will check out on that particular day, but there is significant uncertainty regarding how many will check out between 7 and 7:30 a.m., and how the checkouts will cluster together during this period. As a result, various assumptions regarding the arrival process are required in a queuing model.

Many queuing models presume that clients arrive according to a Poisson process which can be depicted as follows [1].

Let T_i be the *i*th customer's arrival time. Arrivals are said to follow a Poisson process if the subsequent inter-arrival times, namely, $T_1–T_0$, $T_2–T_1$, …, $T_i–T_{i-1}$, …,

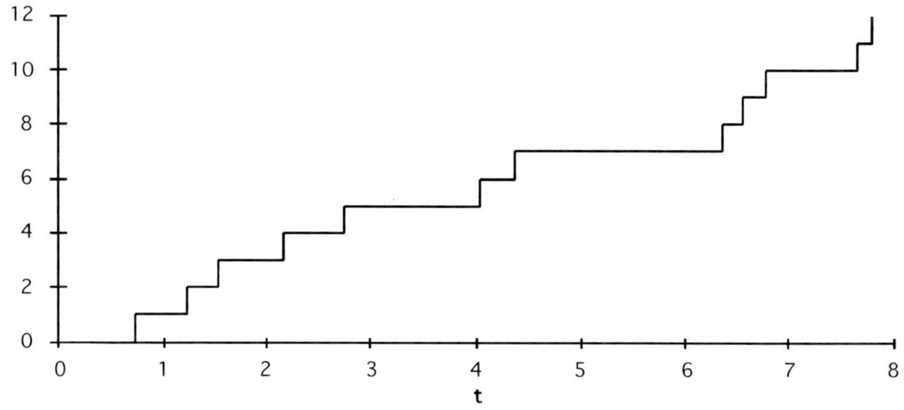

Fig. 5.2 A realization of a Poisson Process adopted from [1]

are independent and all follow the same exponential distribution, i.e., for all t 0 and i = 1, 2,... we have

$$\Pr\,\{T_i - T_{i-1} \leq t\} = 1 - e^{\lambda t}$$

The parameter ***l*****, in this case,** is known as the **arrival rate**, and it represents the average or expected number of arrivals per time unit. Counting arrivals is another method to look at the arrival process. Let $N(t)$ be the total number of arrivals up to time t. $N(t)$ is a random variable that simply counts the number of arrivals for each t. Then $N(t)$—$N(s)$ gives the number of arrivals between time s and time t. (s). Naturally, $N(t) \geq N(s)$ whenever $t \geq s$; therefore, $N(t)$ and $N(s)$ are not independent. Figure 5.2 depicts a typical Poisson process realization.

The Poisson process can be summarized with the following property:

If the arrivals follow a Poisson process, then for every s ≥ t, we can show that $N(t)$—$N(s)$ has a Poisson distribution with mean $l(t{-}s)$. That is, for $k = 0, 1, \ldots$:

$$\Pr\,\{N(t) - N(s) = K\} = \frac{(\lambda(t - s))^k}{k!} - e^{\lambda(t-s)}$$

Furthermore, the quantity of arrivals in any set of non-overlapping time intervals is independent.

Birth–Death Process

The number of persons either waiting in the queue or service, in a queuing system, is said to be the system's state at any time t. The state of the system at time $t = 0$ is the same as the number of people that were present at the beginning. This initial state at $t = 0$ plays a significant role as it impacts the state of the system at any given time t in future.

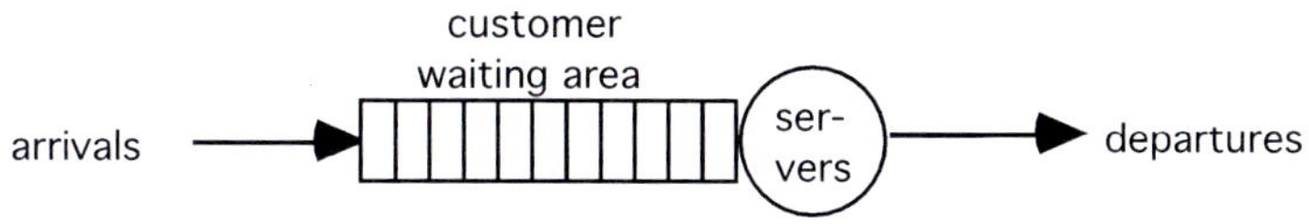

Fig. 5.3 A single-stage queuing system

Figure 5.3 shows the basic (single-stage) queuing system with customers, servers and a waiting queue. In a single-server birth–death process, every birth increases the current state by one. In this system, births are said to occur at a rate λ and deaths at a rate μ. The death decreases the current state by one. λ_j is the birth rate at a state j which represents the possibility of arrival over time. Similarly, μ_j is the death rate at a given state j which represents the probability of service completion over a time. The system is assumed to follow the first-come-first-serve (FCFS) queuing discipline in which customers are served in the order in which they arrive at the system.

Kendall-Lee Notation

The Kendall-Lee notation is a shorthand notation to describe the queuing system. It has 6 abbreviations each representing a characteristic of the system separated by slashes (e.g., *A*/*B*/*C*/*D*/*E*/*F*). The first place (*A* in example) holds the arrival process and the second place (*B* in example) holds the service process. Both A and B are based upon their respective probability distributions.

The possible abbreviations for the first two positions (*A*/*B*) based on the probability distributions are:

M—exponential distribution
E—Erlang distribution
G—General distribution
D—Degenerate distribution (deterministic/fixed service time).

The third place (*C* in example) holds the number of available servers which run parallel to service requests from customers. The fourth place (*D* in example) holds the queuing discipline that will be discussed further in the chapter. The fifth place (*E* in example) holds the maximum limit of customers allowed in the system at any given time. The sixth term (*F* in example) holds the pool of customers that is available, from which the system can draw customers to participate in the servicing.

For example, M/M/3/F CF S/10/ ∞ represents a ticket counter system with exponential arrival and service time distributions. The system has 3 servers (ticket distributing counters) following the FCFS queuing discipline. At any point in time, the system can hold up to 10 customers. However, the system can draw customers from an available infinite population to participate in the distribution.

Little's Formula is given by $L = L_q + L_s$ where L represents the average number of customers in the queue at any given time. L_q is the average number of customers waiting in the queue to get serviced, and L_s is the average number of customers in the queue getting services.

Also, given that W is the average time a customer spends waiting in the queuing system, then we have $L = \lambda W$, with λ being the arrival rate. Thus, little's formula can be further written as $L_q = \lambda\, W_q = > L_s = \lambda\, W_s$, where W_q is the average time spent by a customer waiting in the queue before service and W_s average time spent by a customer waiting in the system getting serviced given that $W = W_q + W_s$.

Queuing Models

Queuing systems are generally of two categories based on the number of servers namely single-server and multi-server queuing systems.

Single-Server Queuing Systems

M/M/1: The M/M/1 system represents a system with an exponential arrival rate and service times with a single server (random arrival, random service and one service channel). This is the most extensively used queueing system in the analysis since almost everything about it is well understood. M/M/1 is a good approximation for many queueing systems. **The M/M/1 results are as follows** [1]:

- W_q (average waiting time for a customer in the queue before service) $= \frac{\rho}{\mu(1-\rho)}$
- W_s (average waiting time for a customer in the system in service) $= \frac{1}{\mu(1-\rho)}$
- L_q (the average number of customers waiting in the queue before service) $= \frac{\rho^2}{1-\rho}$
- L_s (the average number of customers waiting in the system in service) $= \frac{\rho}{1-\rho}$
- The probability of having zero customers in the systems $P_0 = 1-\rho$
- The probability of having n customers in the systems $P_n = \rho_n\, P$

Note: ρ is the utilization factor (the probability/the time that the server is busy.

M/D/1: An M/D/1 queue describes the queue type with a single-server system

(random arrival, deterministic service and one service channel). This system has fixed/deterministic service time and the arrivals are governed by a Poisson process. **The M/D/1 results are as follows:** [1]

- W_q (average waiting time for a customer in the queue before service) $= \frac{\rho}{2\mu(1-\rho)}$
- W_s (average waiting time for a customer in the system in service) $= \frac{2-\rho}{2\mu(1-\rho)}$
- L_q (the average number of customers waiting in the queue before service) $= \frac{\rho^2}{2(1-\rho)}$
- L_s (the average number of customers waiting in the system in service) $= \frac{\rho(2-\rho)}{2(1-\rho)}$

M/G/1: M/G/1 is a queuing model with Markovian arrivals (Poisson process), a general distribution for service times and a single server. For the M/G/1 queue, we have $L_q = \frac{\lambda^2\sigma_s^2+\rho^2}{2(1-\rho)}$ which is called the Pollazcek-Khintichine formula [1].

Similarly for a ***G/G/1*** (a queue following the general distribution of arrival and service rates with a single server) system, the following approximation from [4] is given by

$$L_q \approx \frac{\rho^2\left(1+C_s^2\right)\left(C_a^2+\rho^2 C_s^2\right)}{2(1-\rho)\left(1+\rho^2 C_s^2\right)}$$

In the above equation, if the arrival rate is λ and if σ_a^2 signifies the variance of inter-arrival time, then we have,

C_a^2 (squared coefficient of variation of inter-arrival time) $= \frac{\sigma_a^2}{(1/\lambda)^2}$.

Also, if μ represents the service rate and σ_s^2 represents the variance of the service time, then we have,

C_s^2 (squared coefficient of variation of service time) $= \frac{\sigma_s^2}{(1/\mu)^2}$.

Another powerful approximation of Kraemer and Langenbach-Belz [4] is given by

$$L_q \approx \frac{\rho^2(C_a^2 + C_s^2)}{2(1-\rho)} g$$

where,

$g = \exp\left(\frac{-2(1-\rho)\left(1-C_a^2\right)^2}{3\rho\left(C_a^2+C_s^2\right)}\right)$ when $C_a^2 \leq 1$; and

$g = \exp\left(\frac{(1-\rho)\left(1-C_a^2\right)}{C_a^2+4C_s^2}\right)$ when $C_a^2 > 1$

A few examples to show the application of formulas provided above is shown below.

Example 1: Consider a **single-server** queueing system where the mean inter-arrival time is 20 min and the mean service time is 10 min both **exponentially distributed**.

From the above statement, we can conclude that the system falls under the $M/M/1$ category as the service and arrivals are exponentially distributed (M) with a single server. Also, from the above statement, we can derive the arrival rate (λ) as $\lambda = 1/20$ and the service rate $\mu = 1/10$ and the utilization factor $\rho = 10/20$.

We need to find the following:

i. The average number of customers in the queue: $L_q = \frac{\rho^2}{1-\rho} = \frac{0.5^2}{1-0.5} = 0.5$
ii. The average waiting time in the queue: $W_q = L_q/\lambda = 10$ min.
iii. The average waiting time in the entire system: $W = W_q + 1/\mu = 20$ min
iv. The average number of customers in the entire system: $L = \lambda W = 1$.
v. Probability of the server remaining idle/no customers in the system: $1 - \rho = 0.5$.

Example 2: Consider a single-server queueing system where the mean inter-arrival time is 20 min and the mean service time ranges between 10 and 5 min with a uniform distribution. The arrival time is exponentially distributed.

From the above statement, we conclude that the system is an M/G/1 system as the arrival time is exponentially distributed (M) and the service time is uniformly distributed (G) with a single server. Also, we have an arrival rate $\lambda = 1/20$. To calculate the service time, we first have to average out the range given for the uniform distribution as $(5 + 11)/2 = 8$, which gives $\mu = 1/8$. The service time variance ${\sigma_s}^2$ is calculated as $(11 - 5)^2/12 = 3$ [since any uniform distribution of length 1 has a variance of 1/12]and $\rho = 8/20$.

We need to find the following:

i. The average number of customers in the queue: $L_q = \frac{\lambda^2\sigma_s^2+\rho^2}{2(1-\rho)} = 0.1520$ [for argument sake]
ii. The average waiting time in the queue: $W_q = L_q/\lambda = 3.04$ min
iii. The average waiting time in the system: $W = W_q + 1/\mu = 11.04$ min.
iv. The average number of customers in the system: $L = \lambda W = 0.552$
v. The probability that the system remains idle (no customers in the system): $1 - \rho = 0.6$

Example 3: Consider a **single-server system** where the average arrival time is 20 min and variance is said to be 25 min^2 with a gamma distribution. The mean service time is 10 min with a variance of 20 min^2 with a normal distribution.

From the above statement, we conclude that the system is $G/G/1$ system as the arrival time follows a gamma distribution (G), and the service time is normally distributed (G) with a single server. Also, we have an arrival rate $\lambda = 1/20$ with a variance $\sigma_a^2 = 25$. The service rate $\mu = 1/10$ with a variance $\sigma_s^2 = 20$ and $\rho = 10/20 = 0.5$. Then we have, $C_a^2 = \frac{\sigma_a^2}{(1/\lambda)^2} = 0.0625$ and $C_s^2 = \frac{\sigma_s^2}{(1/\mu)^2} = 0.2$

We need to find the following

i. The average number of customers in the queue: $= L_q \approx \frac{\rho^2(1+C_s^2)(C_a^2+\rho^2C_s^2)}{2(1-\rho)(1+\rho^2C_s^2)} = 0.0322$
ii. The average waiting time in the queue: $W_q = L_q/\lambda = 0.0061$ min ≈ 0.0061 min
iii. The average waiting time in the system: $W = W_q + 1/\mu = 10.0061$ min.
iv. The total number of customers in the system: $L = \lambda W = 0.5$.
v. The probability that the system will remain idle (no customers in the system): $1 - \rho = 0.5$

Multi-server Queuing Systems

M/M/C: Arrivals follow a Poisson process in this model and service times are exponentially distributed. There is a total of s servers ($s > 1$). In the $M/M/C$ model, all arrivals are finally served. This paradigm is simple to understand, and software programs typically provide precise values for the various performance indicators. Here,

$\rho = \frac{\lambda}{C\mu}$ where c represents the number of identical servers

Performance parameters for $M/M/C$ queues are represented as follows: [1]

- The probability of having zero customers in the systems is given by

$$P_0 = \left[\sum_{n=0}^{c-1} \frac{\rho^n}{n!} + \frac{\rho^c}{c!\left(1 - \frac{\rho}{c}\right)}\right]^{-1}$$

- The probability of having n customers in the system is given by

  $P_n = P_0 \frac{\rho^n}{n!}$ for $n < c$ and

  $P_n = P_0 \frac{\rho^n}{c^{n-c}c!}$ for $n > c$

- The average number of customers waiting in the system in service is given by

$$L_s = P_0 \frac{\rho^{c+1}}{\text{cc!}} \frac{1}{\left(1 - \frac{\rho}{c}\right)^2}$$

- The average number of customers waiting in the queue before service L_q is given by

$$L_q = \frac{P_0 \left(\frac{\lambda}{\mu}\right)^c \rho}{c!(1-\rho)^2}$$

- Average waiting time for a customer in the queue before service W_q is given by

$$W_q = L_q / \lambda$$

Similarly, for G/G/C queuing system, we have the approximation as follows: [5, 6]

${W_q}^{G/G/C} \approx {W_q}^{M/M/C} \frac{C_a^2 + C_s^2}{2}$ where ${W_q}^{A/B/C}$ represents the waiting time for a customer in a queue before being serviced in an A/B/C queuing system. The preceding works well for **M/G/c queues**, but not necessarily when the inter-arrival time is not exponentially distributed.

Points to Note:

- A queue with $\rho \geq 1$ and random arrival and/or service rates becomes unstable where the waiting time and queue length tend to infinity. With $\rho > 1$ and constant arrival and service times, the system is unstable and requires additional servers to maintain stability.
- For a uniformly distributed random variable X over the interval (a,b) with a being the minimum and b being the maximum, then the mean of X is given by $(a + b)/2$ and the variance is given by $\frac{(b-a)^2}{12}$. For an exponentially distributed system, the mean is given by $1/\lambda$, and the variance is given $1/\lambda^2$.
- A variable with constant arrival time (deterministic) has 0 variance and coefficient of variance.
- For any two random variables X and Y, we denote mean as $E[.]$ and variance as $V[.]$. Then we have, $E[X + Y] = E[X] + E[Y]$ which implies that if X is the waiting time in the queue and Y is the time taken for service then we get $W = W_q + E[Y] = W_q + \frac{1}{\mu}$.

5.1.2 Delay Differential Equations

We recall the first-order ODE (ordinary differential equation) of the form

$$x^{'}(t) = f(t, x(t))$$

We assume that the derivative $x'(t)$ depends only on the current time t and current state $x(t)$. Here, the system has no dependency on the past. This implies that ODE's cannot directly account for incubation, life history and duration of events. Therefore, ODE's cannot be used to represent anything that had happened in the past and is important for the current evolution of the system. In nature, many changes do not occur instantaneously. For example, considering the birth of human beings, the birth rate depends on the resources available at the point in time. Such instances cannot be captured by the ODEs. Therefore, there arises the need for the DDEs (delay differential equations).

A DDE is an ODE in which the time derivatives depending upon some past values of the state along with the current time, such as,

$x^{'}(t) = f(t, x(t), x(t - \tau))$ where τ represents the lag/delay in the system.

At time t, the evolution of the system depends on the current time t, current state x(t) of the system and at some $\tau > 0$ in the past.

If τ is constant, then the above DDE is a constant DDE.

If $\tau = \tau(t)$ depends on time t, then above DDE is a time dependent DDE.

If $\tau = \tau(x(t))$ depends on state, then the above DDE is state dependent.

A discrete version of DDE with many lags/delays is given as follows,

$$x^{'}(t) = f(t, x(t), x(t - \tau_1) \dots x(t - \tau_n))$$

DDEs can also include derivative of the form,

$$x^{'}(t) = f\left(t, x(t), x(t - \tau_i), x^{'}\left(t - \tau_j\right)\right) \quad i = 1 : n, j = n + 1 : m$$

At time t, the evolution depends on the current time t, current state x(t) of the system and previous state/s of the system x(t-τ) given by,

$x^{'}(t) = f\left(t, x(t), \int_{t-}^{\tau} \phi(x(s))ds\right)$ where ϕ is known as history function.

To solve the ODEs, we set up the initial value problem (IVP) which is accomplished with the initial conditions (IC).

For instance, consider the IVP of the form,

$x^{'(t)} = f(t, x(t)), t \geq t_0$,, the initial condition (IC—the initial value at the initial time point) is given by

$$x(t) = t_0$$

On the other hand, considering a DDE, the IVP is generally of the form,
$x^{'}(t) = f(t, x(t), x(t - \tau_i)) \quad i = 1 : n, t \geq t_0$, for which we construct the IC (initial condition/history function) as

$$x(t) = x_0(t), t \leq t_0$$

Also, for $x^{'}(t) = f(t, x(t), x(t - \tau)), t \geq t_0$ we must provide the initial data on the interval

$$[t_0 - \tau, t_0], \forall t \in [t_0 - \tau, t_0], x(t) = \phi(t)$$

It is to be noted here that even if f is smooth and the same is true for the initial history function $x_0(t)$, we can have.
$\lim_{t \to t_0^-} x_0(\mathrm{t})^{'} \neq \lim_{t \to t_0^+} x_0(\mathrm{t})^{'}$, i.e., there exists a jump derivative discontinuity a t_0.

These discontinuities propagate in time (occurs not only in the initial time point but also at the later time points) which can be shown in the example below as
$x^{'}(t) = -x\left(t - \frac{1}{3}\right) - x\left(t - \frac{1}{5}\right), t \geq 0$ and the history function is given by $x(t) = \delta(t), t \leq 0$, then we can set the initial function to be 0 for all negative values and 1 at initial time point where $t = 0$. We then can obtain the jump discontinuities at the time points 1/3 and 1/5 namely which is depicted in Fig. 5.4.

In the above figure, we can see that for the negative time points the history function $x_0(\mathrm{t})$ is 0 and for $t = 0$ the history function $x_0(\mathrm{t})$ is 1. At points 0.2 and 0.33, we have the jump discontinuities followed by a smoot curve in the later time points.

The dynamical structures of DDEs are much richer that the ODEs. The bounded solutions of autonomous ODEs can only oscillate if there are two components and

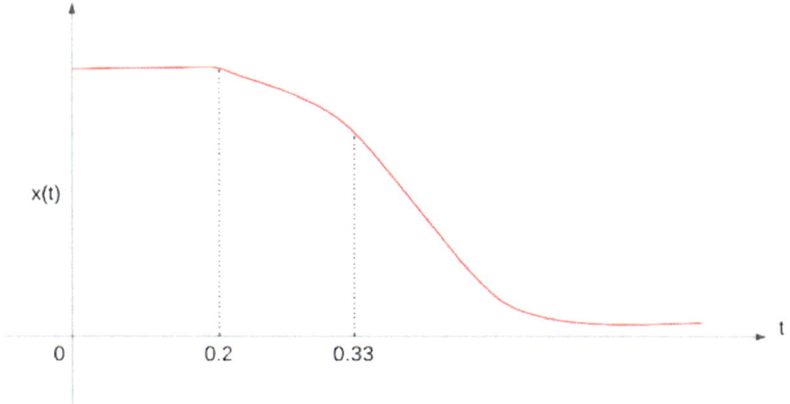

Fig. 5.4 Jump Continuities

we need at least three components in order to have the **chaotic solution.** However, in DDEs both oscillatory and chaotic behaviors can occur in a very simple scalar case.

We will consider the very simple case of DDEs with constant coefficients to explain the various approaches that have been proposed to find an analytical solution to the problem. Here, we consider four such approaches which are.

- Method of steps developed by Bellman and Cooke for variable delays.
- Using Laplace transform methods.
- Matlab solvers for DDE
- Runge–Kutta method

Method of Steps

Method of steps approach starts with the following

i. a constant-coefficient delay differential equation, defined for $t > t_0$
ii. initial function on the interval $[t_0 - \tau, t_0]$, where τ is the delay/lag.

Our interest is to look for uninterrupted progress in future, and hence, we initially consider the interval $[t_0, t_0 + \tau]$ on which the DDE diminishes to ODE. Next, we find a solution on the interval $[t_0, t_0 + \tau]$ and use it as an initial function to find a solution on the next interval $[t_0 + \tau, t_0 + 2\tau]$. In this way, we proceed by progressing the solution from one interval to the next until we arrive at a solution comprising of ODEs that are valid on the interval $[t_0 - \tau, \infty]$. The solution becomes smoother over time t as t increases. At each step, we arrive at a distinctive solution for the ODE by following the hypothesis of the uniform Lipschitz continuity of the right-hand side of equations [7].

Example:

Consider the DDE of the form $x^{'}(t) = x(t-1)$ (constant delay of 1) and the history function.

$$x(t) = 1, \forall t \in [-1, 0]$$

The above equation can be reduced to an ODE on the interval [0,1] as $x^{'}(t) = f(x, x(t))$ with,

$$f(t, x(t)) = \phi_0(t-1)$$

where, $\phi_0(t) = 1, t \in [-1, 0]$

Now making use of integral form of the solutions, we have $\forall t \in [-1, 0]$

$$
\begin{aligned}
x(t) &= x(0) + \int_0^t f(s, x(s))\text{ds} \\
&= x(0) + \int_0^t \phi_0(s-1)\text{ds} \\
&= x(0) + \int_0^t ds \text{ [using } \phi_0(t) = 1, \quad t \in [-1, 0] \text{ as the history function]} \\
&= x(0) + t = 1 + t \text{ as } x(t) = 1, \quad \forall t \in [-1, 0], \\
\text{so} \quad & x(0) = 1
\end{aligned}
$$

Now, continuing the process, we can now write the given DDE on the interval [1, 4] as the non-autonomous ODE as

$$X'(t) = f(t, x(t))$$

With, $f(t, x(t)) = \phi_1(t-1) f(t, x(t)) = \phi_1(t-1)$

Where, $\phi_1(t)$is defined on [0, 1] as $x(t) = 1 + t$. Therefore, the given DDE can be written as,

$x'(t) = \phi_1(t-1), t \in [1, 2]$ where $\phi_1(t) = 1 + t$ for $t \in [0, 1]$.

By making use of the integral form of the solution for $t \in [1, 2]$, we have

$$x(t) = x(1) + \int_1^t \phi_1(s-1)\text{d}s$$

$$= x(1) + \int_0^t \phi_1(s)\text{d}s$$

From solution $x(t) = 1 + t$ as found earlier, $x(1) = 2x(1) = 2$

Therefore, for $t \in [1, 2]$.

Continuing as previously, for $t \in [2, 3]$, we have

$$x(t) = x(2) + \int_2^t \phi_2(s-1)ds$$

Here, the history function $\phi_2(t) = 2 + t + \frac{t^2}{2}$ for t ∈ [1, 4] and $x(2) = 6$. So, for $t \in [2, 3]$

$$x(t) = \frac{10}{3} + 2t + \frac{t^2}{2} + \frac{t^3}{6}$$

Note: Continuing further will help us arrive at a general form of the solution

If instead of constant initial data, that is

$x(t) = 1, \forall t \in [-1, 0]$, we consider the initial data as $x(t) = \xi(t), \forall t \in [-1, 0]$, with $\xi(t)$ function of class C^p on $[-1, 0]$.

Then the solution to the given DDE with the above initial function or the history function is of class C^{p+n+1} on the open interval $(n, n+1)$ for $n \in N$.

To compute the solution on [0,1], we integrate the initial data. So, if initial data on $[-1, 0]$ is $\xi(t) \in C^p$ then the solution on (0,1) is of class C^{p+1}. In turn, we use this solution to construct the solution on (1,2), giving a C^{p+2} solution on (1,2).

Laplace Transform Method

The Laplace transform to a linear constant-coefficient DDE results in an exponential polynomial in the parameter of the transform [8].

Example:

Consider a first-order DDE of the form

$$x' = 2x(t-1)$$

Taking Laplace transform leads to

$$\int_1^\infty x'(t)e^{-st}\mathrm{d}t = 2\int_1^\infty x(t-1)e^{-st}\mathrm{d}t$$

From this, we have (using change of variable on the right-hand side)

$$\left[x(t)e^{-st}\right]_1^\infty \int_1^\infty x(t)\left(-se^{-st}\right)\mathrm{d}t = 2\int_0^\infty x(u)e^{-s(1+u)}$$

Furthermore, assume that $x(t)e^{-st} \to 0$ and $t \to \infty$, leads to

$$-e^{-s}x(1) + s\int_1^\infty e^{-st}x(t)\mathrm{d}t = 2e^s\int_0^1 x(u)e^{-su}\mathrm{d}u + 2e^{-s}\int_1^\infty x(u)e^{-su}\mathrm{d}u$$

Also, assume that $\int_1^\infty x(t)e^{-st}\mathrm{d}t = \frac{x(t)e^{-s}+2e^{-s}\int_0^1 x(u)e^{-su}\mathrm{d}u}{s-2e^{-s}}$

Assume that the inversion formula can be applied to have

$$x(t) = \int_C \frac{x(t)e^{-s} + 2e^{-s}\int_0^1 x(u)e^{-su}\mathrm{d}u}{s-2e^{-s}}e^{st}\mathrm{d}s$$

Hence, given that all the steps can be justified, the solution can be expressed in terms of initial values of $x(t)$ over [0, 1] employing a contour integral.

Matlab Solvers for DDE

The following table shows the available solvers for the various types of delay differential equations (DDE) [7] and the same is shown in Fig. 5.5.

S. No.	Solvers	DDE's
1	dde23	To solve DDEs with constant delays
2	ddesd	To solve DDEs with general delays
3	ddensd	To solve DDEs of neutral type

To explain the usage of Matlab, we choose one of the solvers namely dde23 for demonstration as shown in Fig. 5.6. We choose a special case 'DDE with constant delays'.

The following figures show an example of using the DDE in Matlab. Figure 5.7 shows the documentation/guide page and an example to use dde23 solver for coding DDE with constant delays.

Figure 5.8 shows the dde23 editor with a system of equations along with the history function which is to be coded in the dde23 solver format. The system of equations is also shown in Fig. 5.8.

For $t \leq 0$, the history function is constant, $y_1(t) = y_2(t) = y_3(t) = 1$. The equations represent a system of constant delay equations as the temporal delays are present only in the y terms which are in turn constants.

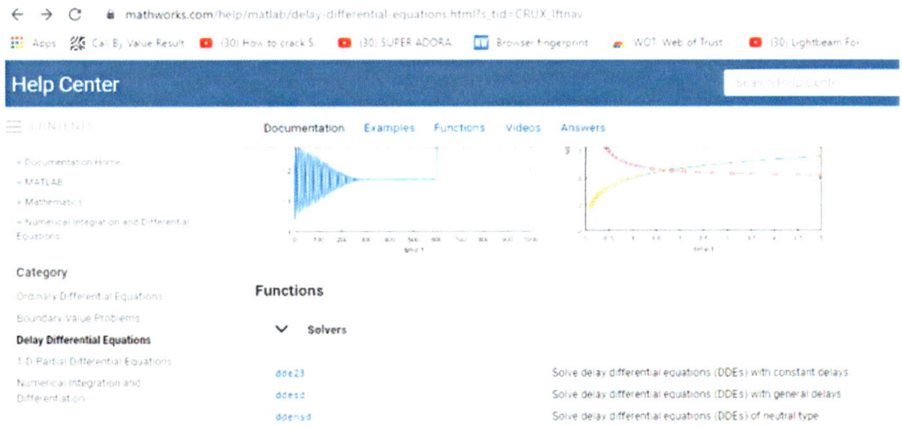

Fig. 5.5 Screenshot of available Matlab solvers

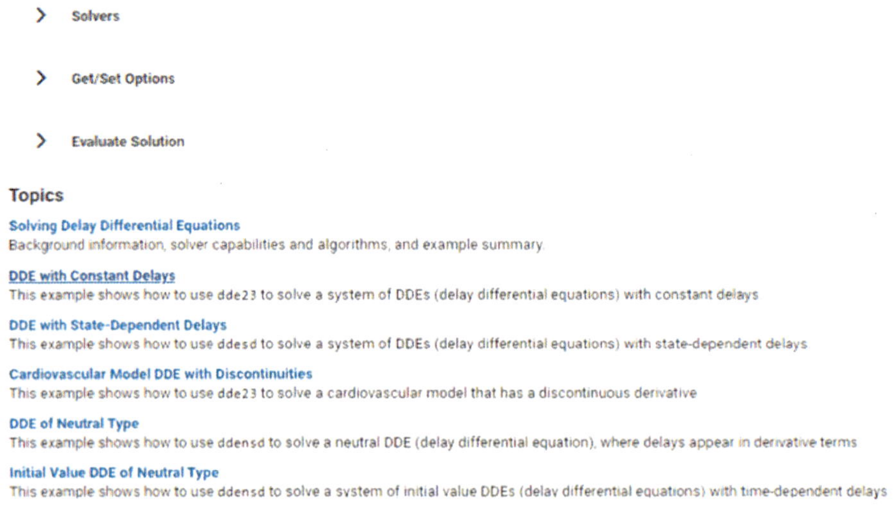

Fig. 5.6 DDE with constant delays clickable link

Code Delays

First, the system of equations shown in Fig. 5.8 needs to be coded before calling the delay differential equation solver dde3. As shown in Fig. 5.9 under the 'Code Delays' segment, we begin by defining the delays in the system of equations with a vector. There are two types of delays in this system:

- Component 1: $y_1(t-1)$ with a delay of 1.
- Component 2: $y_2\,(t-0.2)$ with a delay of 0.2.

For the delays, dde23 accepts a vector parameter, with each element representing the constant delay for one component.

```
lags = [10.2]
```

Code Equations

Next, we code the system of equations mentioned above into a function as shown in Fig. 5.10.

This function's signature should be $dydt = ddefun(t, y, Z)$, where

- Time is given by t (independent variable).
- The solution is given by y (dependent variable).
- $Z(:, j)$ approximates the delay $y\left(t-\tau_j\right)$, where lags(j) denotes the constant delay $y\left(t-\tau_j\right)$.

The solver automatically passes these inputs to the function, but the variable names influence how we code the equations. The function is represented in Fig. 5.10.

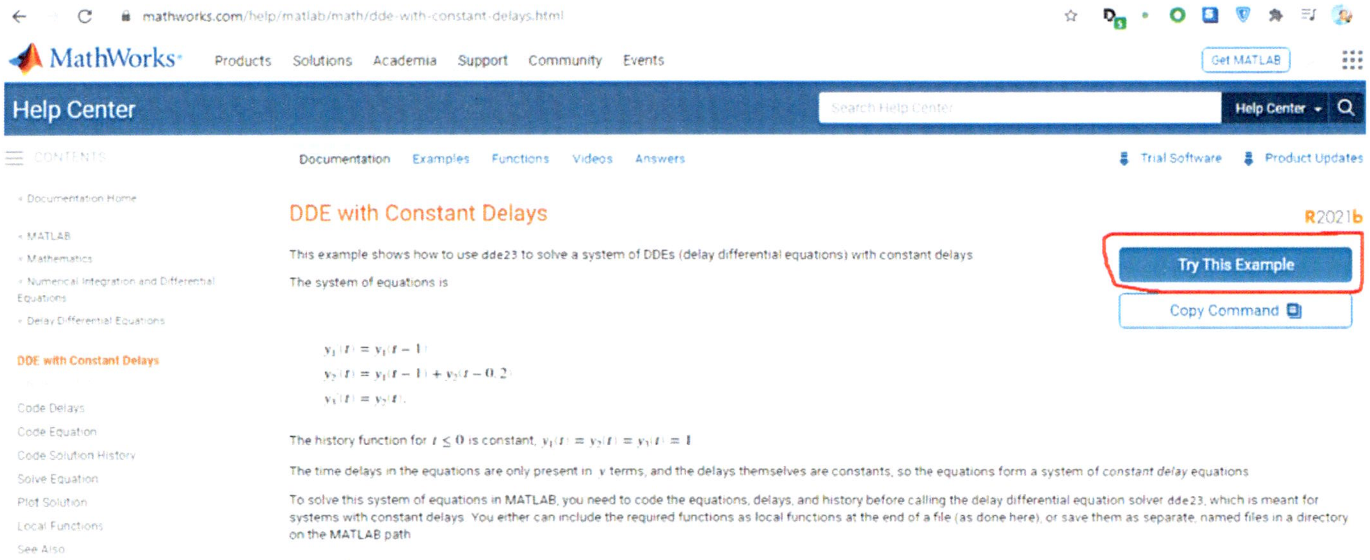

Fig. 5.7 Documentation/guide to using dde23 solver for coding DDE with constant delays

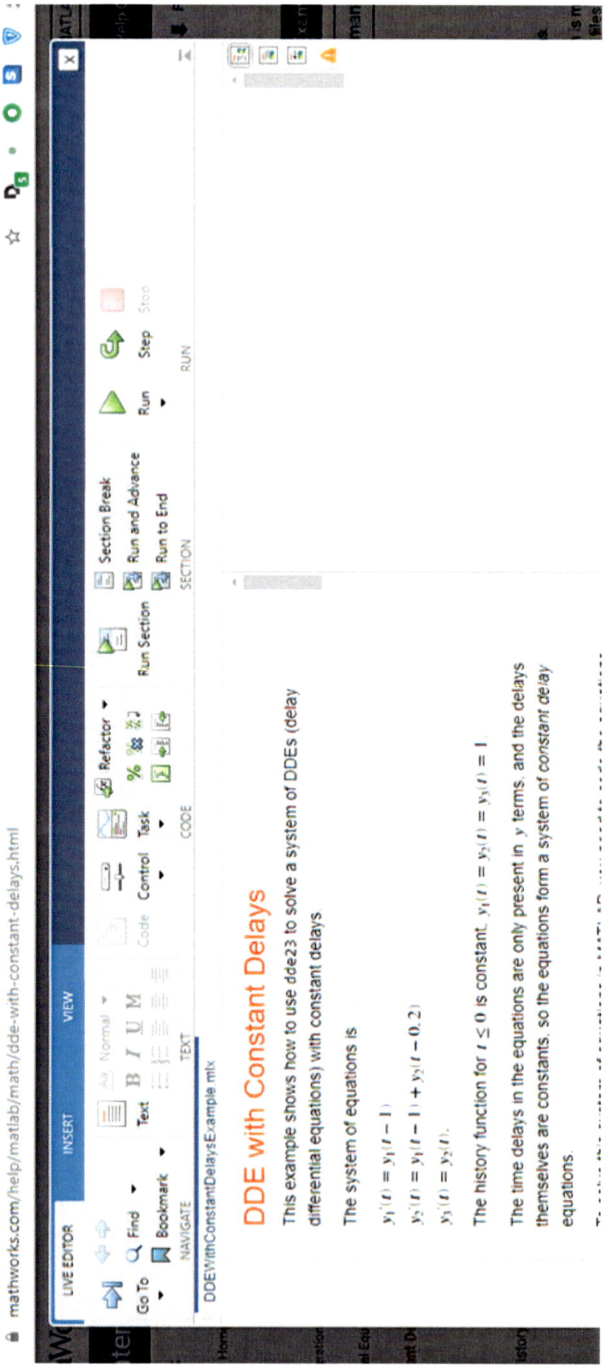

Fig. 5.8 A system of DDE equations along with the history function

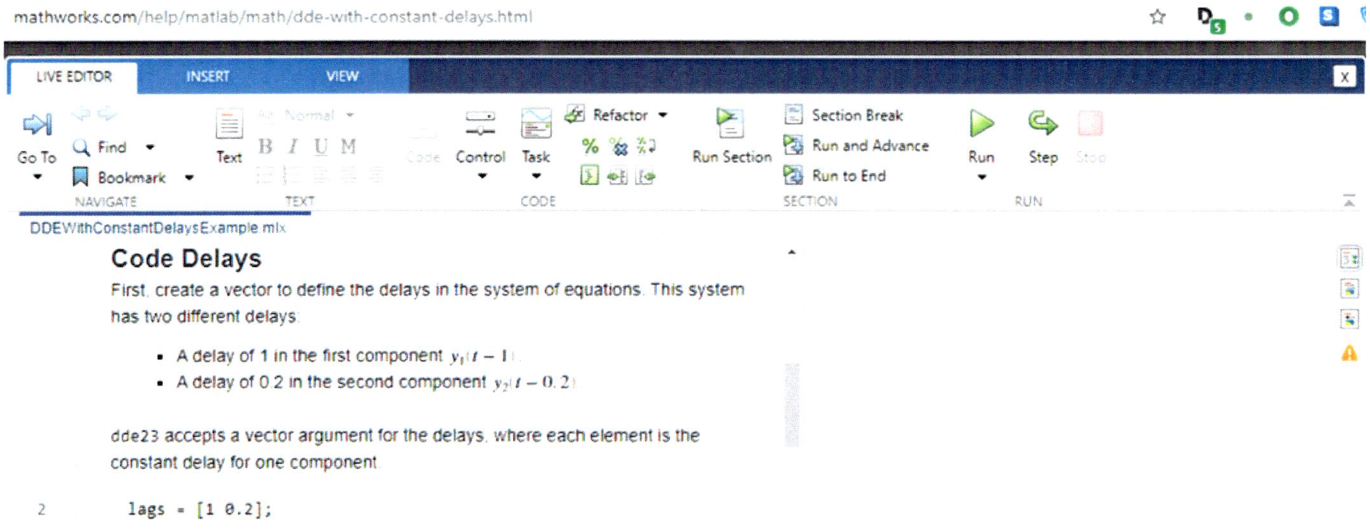

Fig. 5.9 Code the delays

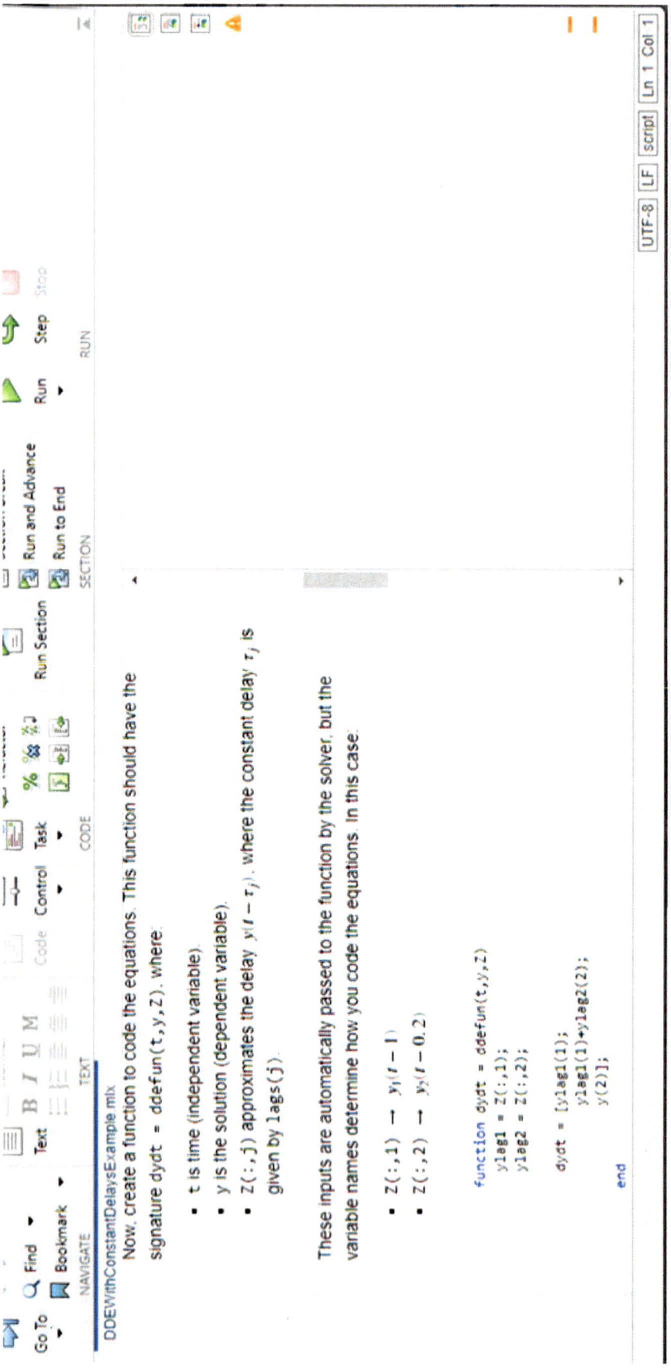

Fig. 5.10 Coding the system of equations

Code Solution History

Next, we develop a function that defines the solution history as shown in Fig. 5.11. The solution history is the solution that spans for the time stamps $t \leq t_0$.

The code for the solution history is as follows [9]:

```
Function S = history (t)
S = ones (3,1)
End
```

Solve Equation

The integration interval $\left[t_0 t_f\right]$ to solve the DDE using the dde3 solver needs to be coded as shown in the 'Solve Equation' section in Fig. 5.11. The code is shown in Fig. 5.11.

Plot Solution.

The solution structure 'sol' has the fields sol.x and sol.y which hold the solver's internal time steps and related solutions at those times. We plot the solution component against time as shown in Fig. 5.12.

Figure 5.13 shows the entire code (local helper function) that the dde3 solver calls to solve the system of equations, along with the final output obtained upon running the dde3 solver on the right-hand side of the window.

The Runge–Kutta Method for Delay Differential Equations:

Apart from the above shown methods, we can adapt many other numerical methods such as the Euler's method, multistep and block implicit methods that are used to solve the ordinary differential equation (ODE) of the form

$$y^{'(t)} = f(t, y(t)) \quad y(t_0) = c \tag{5.1}$$

to solve the delay differential equations. Here, we quote [10] to demonstrate the adoption of the Runge–Kutta method for solving delay differential equations. When solving Eq. (1) at the point t_{n+1} using a q-stage Singly Diagonally Implicit Runge–Kutta (SDIRK) technique, the following equations are obtained:

$$k_1 = f(t_n + c_1 h, y_n + h a_{11} k_1)$$

$k_2 = f(t_n + c_2 h, y_n + h a_{21} k_1 + h a_{22} k_2)$ and so on which leads to a general form as given below

$$k_i = f\left(t_n + c_i h, y_n + h \sum_{j=1}^{i} a_{ij} k_j\right)$$

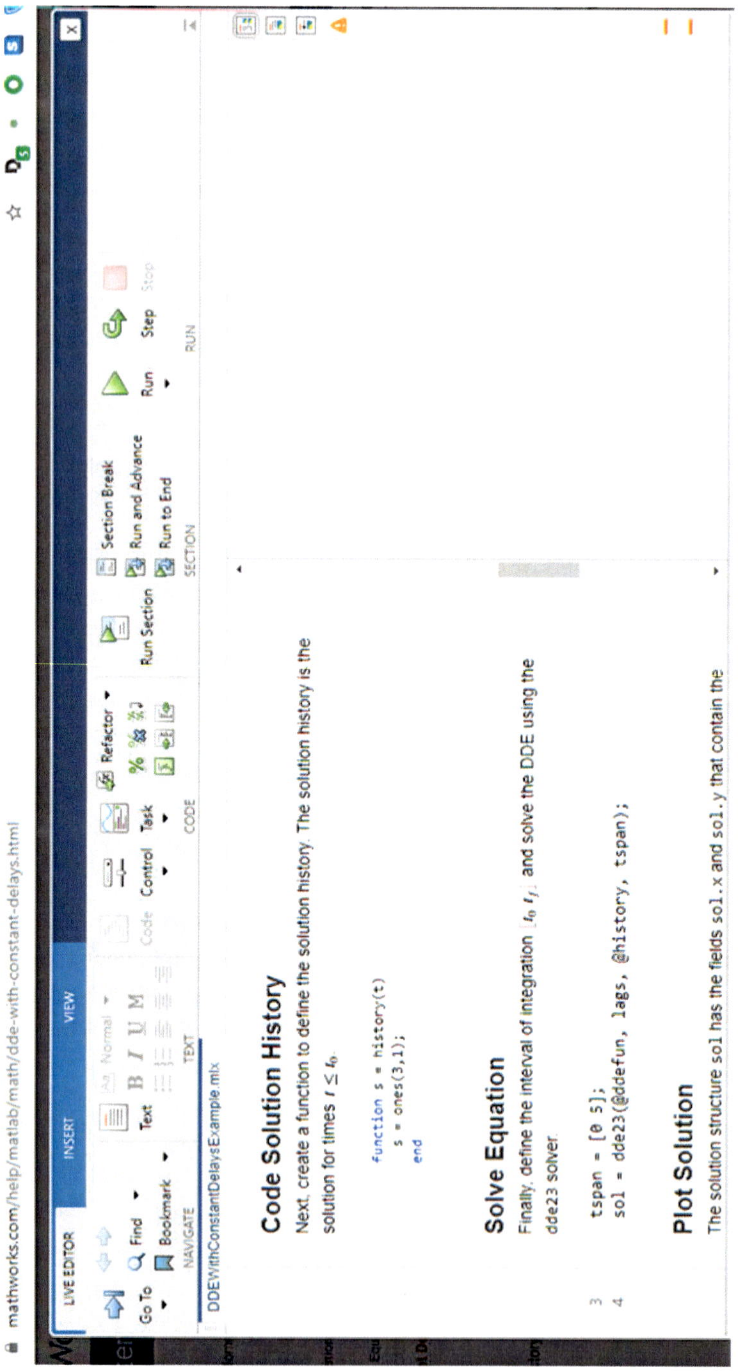

Fig. 5.11 Code the History function (Code solution history) and defining the integration interval (solve equation) for the dde3 solver

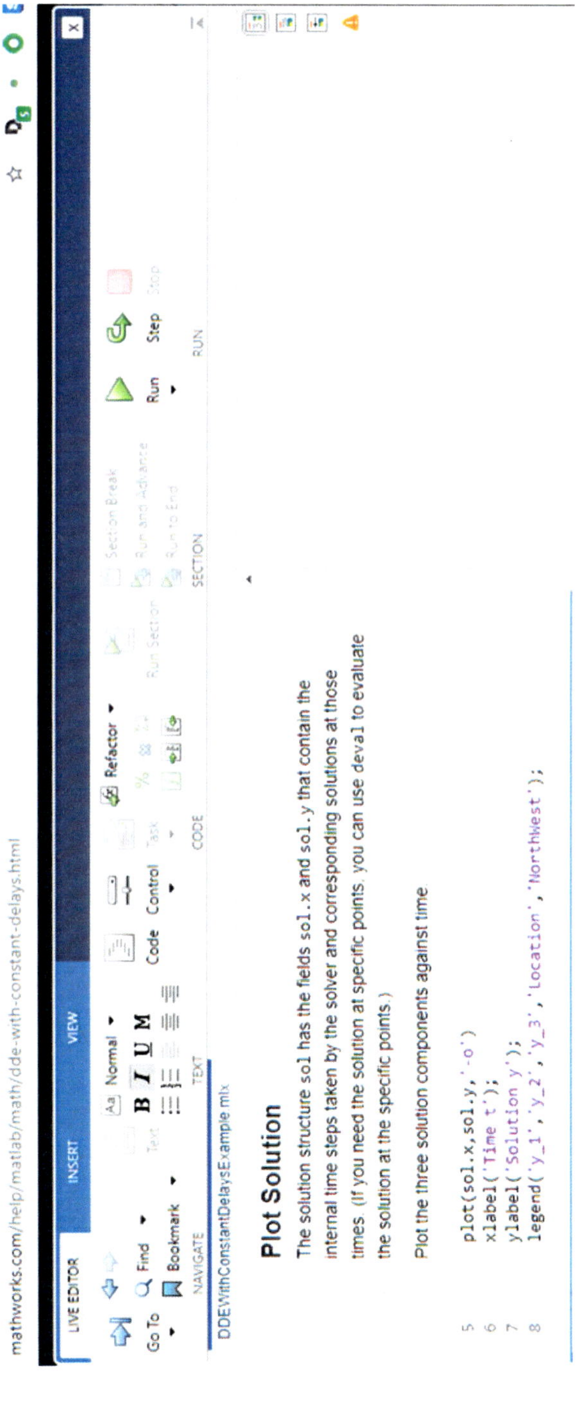

Fig. 5.12 Specifying the solution components (y_1, y_2, y_3) against time

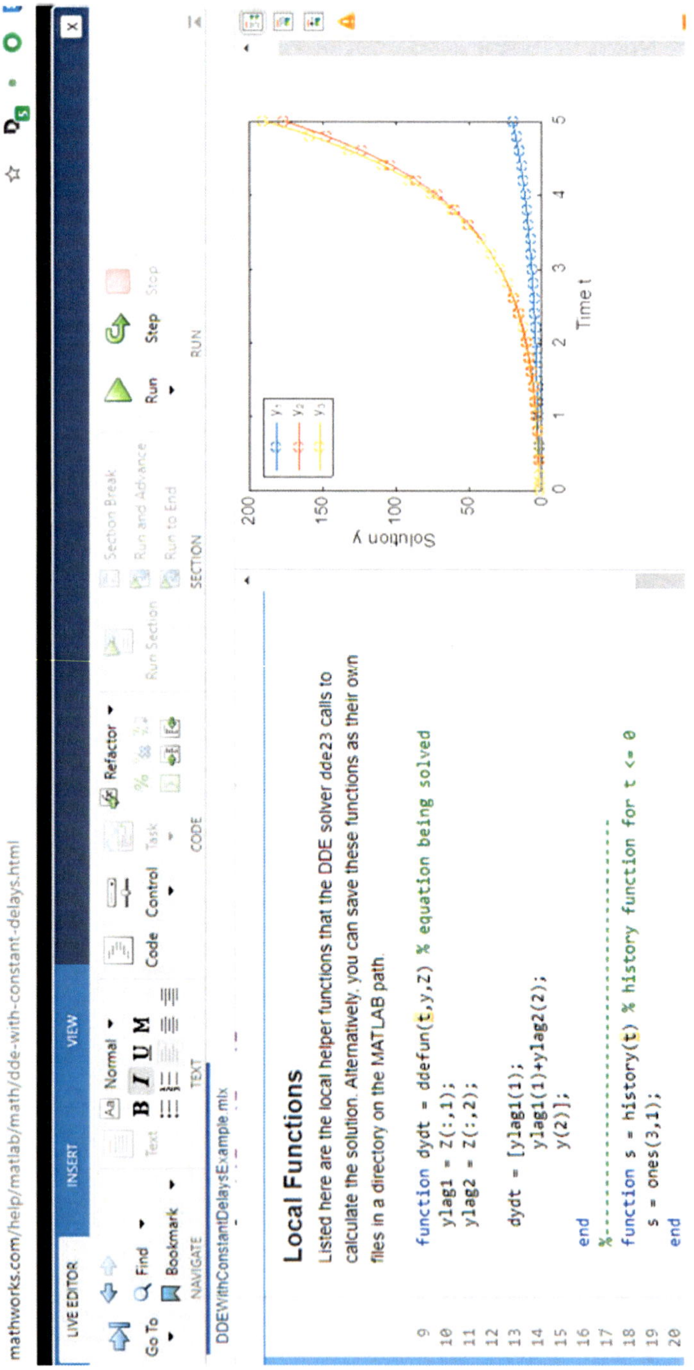

Fig. 5.13 All function combined under one roof along with the solution on the right-hand side

$$y_{n+1} = y_n + h \sum_{i=1}^{q} b_i k_i (i = 1, \ldots, q)$$

Here, $y_n + h \sum_{j=1}^{i} a_{ij} k_j$ is the integral stage of the method. When we extend the above equations to DDE, we get,

$$k_i = f\left(t_n + c_i h, y_n + h \sum_{j=1}^{i} a_{ij} k_j, y(t_n + c_i h - \tau)\right)$$

$$y_{n+1} = y_n + h \sum_{j=1}^{i} b_i k_i$$

Here, the delay term is $y(t_n + c_i h - \tau)$ and interpolation is required to estimate the value. Hermite interpolation is employed to estimate the delay term in 'Numerical treatment of delay differential equations (DDE) by Runge–Kutta method using Hermite interpolation'. The interpolation order, and thus the number of support points, must be suited to the method's order. If p represents the order of the Runge–Kutta method, the interpolation order q must be greater than or equal to p. If i_p specifies the number of Hermite interpolation support points, then

$$2i_p > p$$

Applications of Delay Differential Equations

The delay differential equations play a vital role in mathematically modeling many practical applications across various disciplines such as engineering, control theory and biosciences. All systems that have a feedback mechanism will always have time delays since the systems take a finite time to sense information and respond to it. The delays may include gestation times, transport delays and so on. Delay models also play a vital role in the medical diagnosis systems in describing various facts about the diseases and dynamics such as primary infection, immune response and drug therapy. They also find their application in the study of epidemiology, tumor growth, cardiac rhythms and so on. Delays have also been identified in the population and evolution dynamics of many species identified by conducting statistical analysis on ecological data. Delay differential equations are applied to mathematically model various physical and biological aspects such as disease spread and human pupil light reflex. They have also been applied in the field of control theory to depict real-life simulations of examined occurrences. DDE s have been used to monitor the combustion in the chamber of turbo engines which is an example of its application in the fields of mechanical engineering.

5.2 Motivation

Dynamical systems with time delays have recently gained popularity. Their applications span from biology and population dynamics to physics and engineering, and from economics to medicine. So, what is the significance of time-delayed systems? The straightforward explanation is that time delays are inherent in many real-world systems and must be adequately accounted for when creating models. Many genuine processes involve delays, and as the demand for more exact forecasts, control and performance grows, so does the requirement for models to behave as closely as feasible to real systems. Time delays are common in feedback loops involving sensors and actuators in engineering. The transmission of measured signals to a remote-control center is becoming easier as communication technologies advance. However, engineers face a substantial challenge in the form of an inevitable time delay between the measurement and the signal received by the controller, which should be factored in at the design stage to avoid experimental instability and probable damage.

5.3 Case Studies

This section explains the application of the mathematics that has been discussed in the previous sections to some real-world problems. The usage of the two categories namely queuing theory and delay differential equations in modeling a system/problem is explained with the help of two specific case studies adopted by the authors in their research papers.

Case Study 1: Computer Engineering Application: Future Internet (Queuing Theory)

We explain the usage of queuing theory in modeling the system by quoting the research problem from [11]. In [11], the authors have proposed a system that is resilient to various threats and attacks posed to the cloud system in future Internet scenario. The architecture of the proposed system is shown in Fig. 5.14. The system is perceived to be an extension of the M/D/1 queuing system with multiple server where the arrivals are exponential and the service times are deterministic. The communication between the component in the system is categorized and modeled under three different scenarios. The authors have derived an equation to depict the total time spent by the devices in the system under the three scenarios represented in the Fig. 5.14 as $E[T] = E[T_{\mathrm{KDC}}] + E[T_{\mathrm{AA}}] + E[T_{\mathrm{Cloud}}]$ where $E[T_{\mathrm{KDC}}]$ represents the total time spent by devices in the subsystem labeled as scenario 1. $E[T_{\mathrm{AA}}]$ represents the total time spent by devices in the subsystem labeled as scenario 2, and $E[T_{\mathrm{Cloud}}]$ represents the total time spent by devices in the subsystem labeled as scenario 3. Scenario 1, as shown in Fig. 5.15, follows a M/D/1 queuing model as the communication between devices that arrive at an exponential rate sends requests to AA/KDC (key distribution center) which a single server for the domain. Approaching the subsystem in this way, we arrive at a solution $E[T_{\text{KDC}}] = E[S_{\text{KDC}}]\left[\frac{1+\rho_{\text{KDC}}}{2(1-\rho_{\text{KDC}})}\right]$.

Scenario 2 as shown in Fig. 5.16, follows an M/D/n queuing model which is a novel contribution by the authors that depicts the communication between the cloud and the AA (multiple servers) which helps us arrive at an equation for the total time spend in the subsystem labeled scenario 2 as $E[T_{AA}] = \sum_{i=1}^{n}\left(E[S_i]\left(1 + \frac{\rho_{AA_i}}{2(1-\rho_{AA_i})}\right)\right)$.

Scenario 3, as shown in Fig. 5.17, follows an M/D/1 queuing model as it depicts the communication between the devices and the cloud (single server) which leads us to the equation for the total time spent by the device in the subsystem labeled scenario 3 as $E[T_{\text{Cloud}}] = E[S_{\text{Cloud}}]\left[\frac{1+\rho_{\text{Cloud}}}{2(1-\rho_{\text{Cloud}})}\right]$. Finally, the total time spent in the entire system is given by $E[T] = \sum_{i=1}^{n}\left(E[S_i]\left(1 + \frac{\rho_{AA_i}}{2(1-\rho_{AA_i})}\right)\right) + E[S_{\text{Cloud}}]\left[\frac{1+\rho_{\text{Cloud}}}{2(1-\rho_{\text{Cloud}})}\right]$.

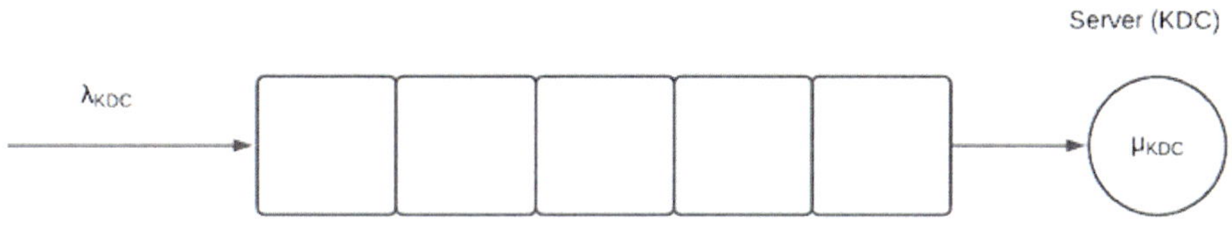

Fig. 5.14 M/D/1 Queuing system for scenario 1 (KDC) adopted from [11]

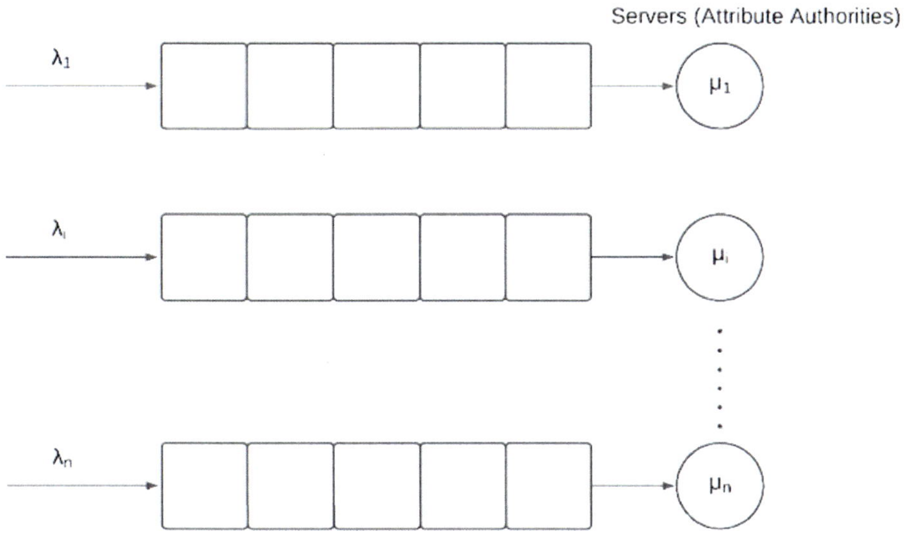

Fig. 5.15 M/D/n Queuing System for Scenario 2(AA) adopted from [11]

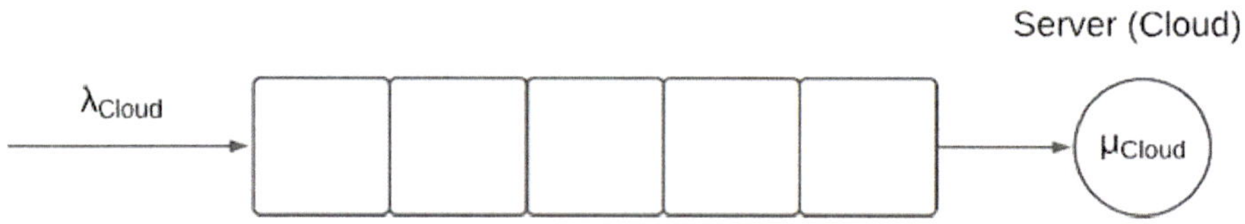

Fig. 5.16 M/D/1 Queuing system for scenario 3(Cloud) adopted from [11]

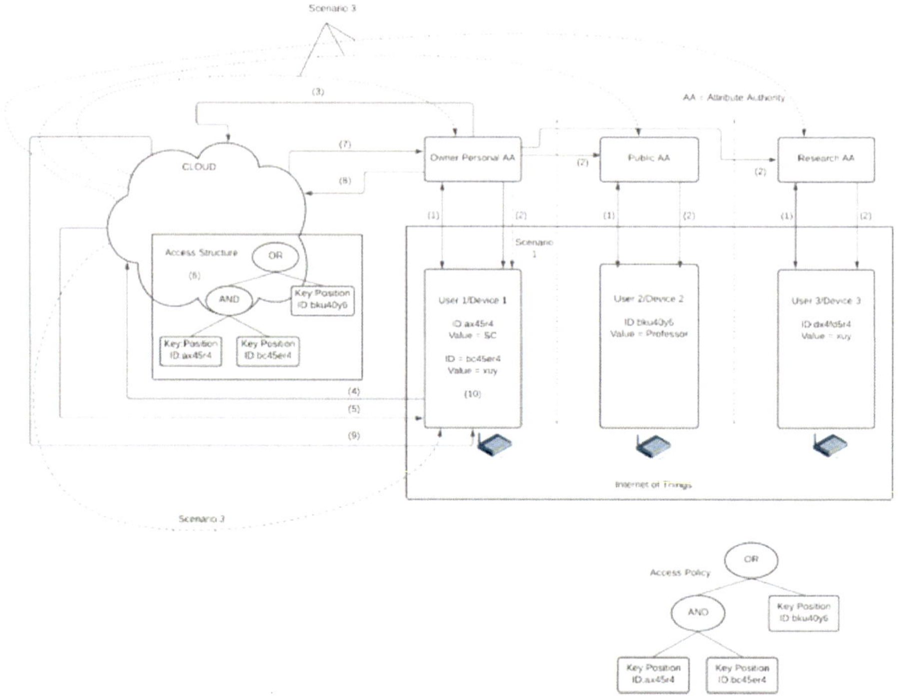

Fig. 5.17 Proposed IAMFI Architecture adopted from [11]

Case Study 2: Mechanical engineering application: metal rolling system (delay differential equation)

We explain the usage of delay differential equations in modeling the system by quoting the metal rolling system [12] which is a mechanical engineering application. Ribbons of heated metal sheets pass between rollers in a metal rolling system, as seen in Fig. 5.18.

The purpose is to regulate the upper roller to generate metal sheets of a specific thickness, whereas the lower one is set in stone. This is powered by a drive motor. The thickness of the sensor is placed downstream at a distance d between the rollers. The ribbon is fed into the machine at a speed v. The delay is dependent on both the distance and speed, which is given by $\tau = d/v$. The measured thickness is denoted by $y(t-\tau)$, the actual thickness by $y(t)$ and the desired thickness by $y_d(t)$. The DDE is obtained as follows

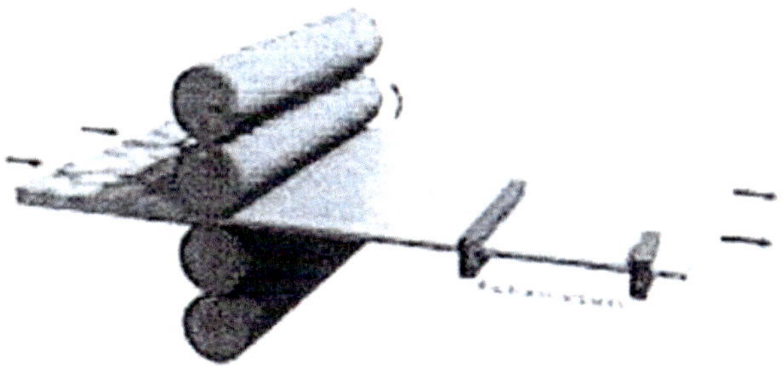

Fig. 5.18 Metal rolling system [12]

$$y'(t) + k_p k_m y(t - \tau) = k_p k_m y_d(t)$$

where k_p and k_m are proportional controller gains in the above equation. The metal rolling system's state-space representation is given below

$$\begin{cases} x'(t) = A_d x(t - \tau) + Bu(t), \quad x(0) = x_0 \\ y(t) = x(t) \end{cases}$$

Figure 5.19 depicts the block diagram of this metal rolling mechanism in the above equation. Suppose that the transfer function between the top roller position and the driver motor's input voltage is $G(s) = k_m/s$. The controller's transfer function is $C(s) = k_p/s$. The censor's transfer function is $H(s) = e^{-\tau s}$. $P(s)$ denotes the effect of external disturbances on the system. Presume that $G(s)$ is provided by

$$G(s) = \frac{k_m}{s(1 + Ts)}$$

We finally obtain the second-order delay differential equation (DDE) as follows

$$y''(t) + y'(t) + k_p k_m y(t - \tau) = k_p k_m y_d(t)$$

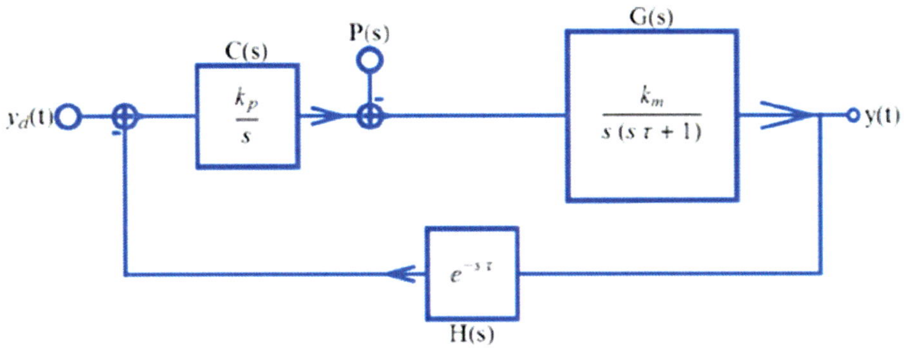

Fig. 5.19 Block diagram of the metal rolling system adopted from [13]

5.4 Summary

This chapter has extracted and presented a few mathematical concepts that are suitable for the representation and analysis of delay in engineering problems across domains. This will provide an understanding to the interested readers, thereby making their life easier in representing the problem at hand mathematically. Identification of the appropriate mathematical concept to aptly represent the problem is the goal that has been achieved in this chapter. The examples and case studies to show the research applications of the explained concepts are bonuses that will help researchers in understanding the concepts better and find a connection to their problem, thereby making it easier to apply with a few tweaks. The major challenge in presenting the chapter was the election of the appropriate mathematical concepts from the vast domain. We have followed an introductory approach of the whole subject which is sufficient for researchers to brush up the concepts. Major focus has been directed toward the application of the concepts to research problems and develop a confidence among the interested readers.

References

1. Sheldon M (2014) Ross, introduction to probability models, 11th edn, Elsevier Science
2. Tornambe (1995) Elementary queuing theory. Discrete event systems theory an introduction
3. Gross D, Shortie JF, Thompson JM. Carl MH (2008) Fundamentals of queuing theory. Willey
4. Marchal WG (1976) An approximate formula for waiting time in single server queues. A I I E Trans 8(4):473–474. https://doi.org/10.1080/05695557608975111
5. Medhi J (2003) Stochastic models in queueing theory, 2nd edn. Academic Press, California, USA
6. Whitt W (1974) Heavy traffic limit theorems for queues: a survey. Springer-Verlaq, Berlin, Heidelberg, New York, pp 1–46
7. Jumma SI (2017) Solving linear first order delay differential equations by MOC and method of steps comparing with matlab solver, Thesis, NEU, Nicosia
8. Cimen E, Uncu S (2020) On the solution of the delay differential equation via laplace transform. Commun Math Appl 11(3):379–387. https://doi.org/10.26713/cma.v11i3.1359
9. https://www.mathworks.com/help/matlab/math/dde-with-constant-delays.html
10. Ismail F, Al-Khasawneh RA, Suleiman M (2002) Numerical treatment of delay differential equations by Runge-Kutta method using Hermite interpolation. Matematika 18
11. Nancy Ambritta P, Railkar PN, Mahalle PN (2014) A queuing theory-based modelling for performance analysis towards future internet. In: 2014 Annual IEEE india conference (INDICON), pp 1–7. https://doi.org/10.1109/INDICON.2014.7030521
12. Mirkin L (2012) Introduction to time-delay systems. Lectures of automatic control–doctorate program. Lund Univ.(2012). See http://www.control.lth.se/media/Education/DoctorateProgram/2012/Lectures
13. Boukas E-K, Liu Z-K (2002) Deterministic and stochastic time delay systems. MA, USA, Birkhäuser, Boston
14. Kershenbaum A (1993) Telecommunication network design algorithms. McGraw Hill

Chapter 6
Data Science Problems

6.1 Introduction

Data are elements or information that are usually numerical and are collected by observation. It can also be defined as a set of values (quality or quality) related to people or things.

Data science is the process of extracting important **information** from collected data**.** It is a multidisciplinary area that uses various algorithms and scientific methods to extract information from data from various sources that may be structured or unstructured. The knowledge that is acquired in the process is then used across various application areas to expose various dimensions. A strong foundation in the required mathematical concepts combined with the domain knowledge along with the required IT/programming skills can help in putting data science into action to achieve greater goals.

Data science includes the use of computer science, mathematics, machine learning, visualization and human–computer interaction to collect, clean, compile, analyze, visualize and interact with data collected to create data products. A **data set** is a collection of data objects or businesses (such as clients, patients, students, professionals, etc.). Each data item has **attributes** that describe objects (for example, characteristics that identify a person (object) can include age, gender, suitability, etc.).

Mathematics is a very important tool that helps in the analysis of this data. Data science cannot exist without the underlying mathematical tools. Mathematical tools like probability, differential equations, linear algebra, optimization, graph theory and calculus are a few areas of interest that play a role in laying down the foundation for data science.

Statistics help to better understand the data which includes important metrics namely central tendency(average), variance and spread [1]. Data distribution characteristics are measured in metrics such as median, max, min, quartiles, outliers, variance, etc. A few important mathematical aspects of data processing and utilization are discussed further in this chapter.

P. N. Mahalle et al., *Foundations of Mathematical Modelling for Engineering Problem Solving*, Studies in Autonomic, Data-driven and Industrial Computing,
https://doi.org/10.1007/978-981-19-8828-8_6

Measures of Central Tendency

Measures of central tendency assists in providing statistical explanation to data based on an average value from the database. They are also called summarized statistics. They include the mean, median and mode.

Mean (value) represents the sum of the whole set of values in a data set divided by the total number of values that exist in the data set. Since 'n' refers to the total number of entries in the data set (sample size) and × 1, × 2, × 3..., refers to the individual values in the database, the **sample mean** ($\overline{x}$) is given as.

$\overline{x} = \frac{x_1+x_2+x_3+...x_n}{n} = \frac{\sum x}{n}$. The **population mean** (μ) is given by $\mu = \frac{\sum x}{N}$ where 'N' is the population size. The **weighted arithmetic mean** ($\overline{x}$) is given by $\overline{x} = \frac{\sum_{i=1}^{n} w_i x_i}{\sum_{i=1}^{n} w_i}$ where 'n' is the sample size and 'w' refers to the weights.

The **median** is determined to be the middle value when positioned in order of magnitude (ascending order). For example, for a data set as given below **(odd number of values),**

34	65	22	78	54	12	23

We first rearrange the data set in ascending order as

12	22	23	**34**	54	65	78

The median value after the reset in this case (odd number) is the value at the 4th position '34'.

In the case of a database with equal values, we take an average of two median values to provide the median. For example,

12	22	23	34	54	65	78	80

In the above case (even), we have 34 and 54 as the middle values. Hence, the median is given as $\frac{34+54}{2} = 44$. The median is least impacted by outliers or distorted/skewed data in the database.

Mode is the most common value in the database. It describes the most widely used or most common option in the population. The empirical value is given by mean-mode = 3 (mean-median).

Symmetric and **skewed data** can be better defined with the help of the graphs given below. A symmetric data set has a normal distribution. This is where the median and the mode intersect in the same area as shown in Fig. 6.1. The distorted/skewed data is when the frequency curve of the observations is inclined to the right (**negatively skewed** i.e., mode > median) or left (**positively skewed** i.e., mean > median) as shown in Figs. 6.2 and 6.3.

The use of various approaches of calculating the central tendency may differ subject to the type of data. However, there is no specific rule for its use. The mean is the most appropriate measure of central tendency for a data set with continuous and consistent data (data is normally distributed).

The least common measure is the mode and is usually applied for nominal data. While operating with other types of data, mean or median are commonly desired, but this does not mean that they are never used with these types of data. A median is generally employed for data that is skewed or ordinal data.

Measures of dispersion in statistics indicate the skewness or distributed nature of the data. It helps in interpreting the data variability/spread. It points out the homogeneity/heterogeneity of the data.

Quartiles are values that divide the data set into quarters given as Q_1 (25th percentile), Q_3 (75th percentile) and so on.

Inter-quartile range is the difference of the values Q_3 and Q_1 which is given by

$$\mathbf{IQR} = \boldsymbol{Q_3 - Q_1}$$

Five number summaries help us to obtain a rough summary of the actual data with the five datapoints namely the **min,** $\boldsymbol{Q_1}$**, median,** $\boldsymbol{Q_3}$ **and max**.

A **boxplot** is a graphical representation of the data that indicates the spread, locality and skewedness. The ends of the box are the quartiles, median is marked and whiskers are added followed by the outliers that are marked individually as shown in the Fig. 6.4.

Any significantly different data points from the rest of the values are called **outliers** and are typically a value greater or lesser than **1.5 x IQR.**

The **variance** of a set depicts how further away is each number from the mean and from every other number in the set. **Standard deviation** is the square root of the variance.

Given that 's' is the sample size and σ is the population, the variance and standard deviation is given by.

Fig. 6.1 Symmetric data

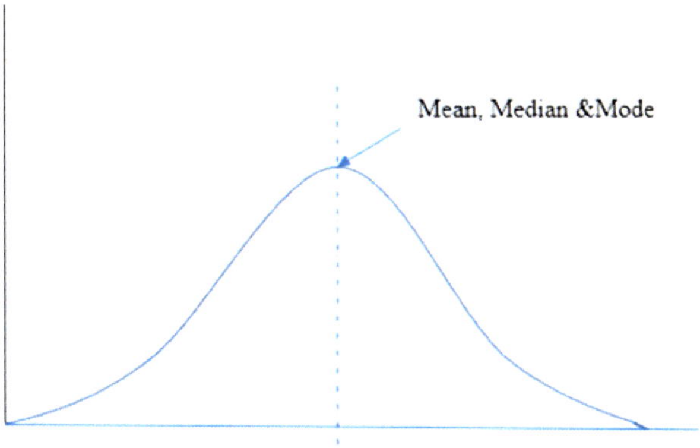

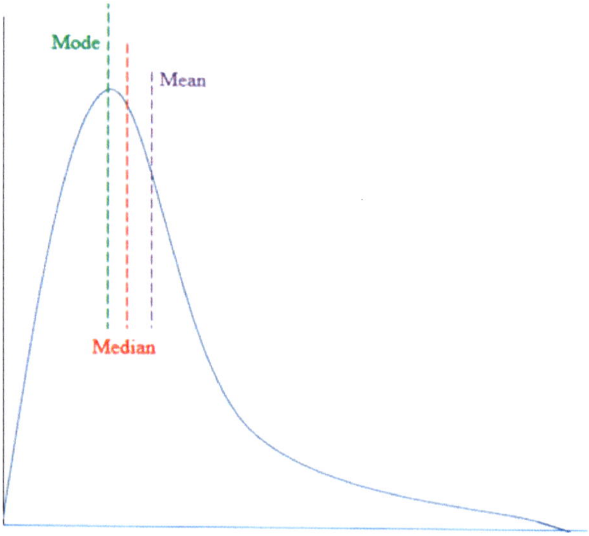

Fig. 6.2 Positively skewed data

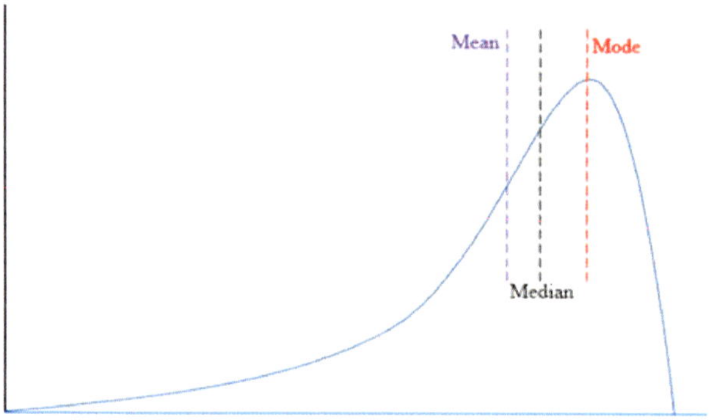

Fig. 6.3 Negatively skewed data

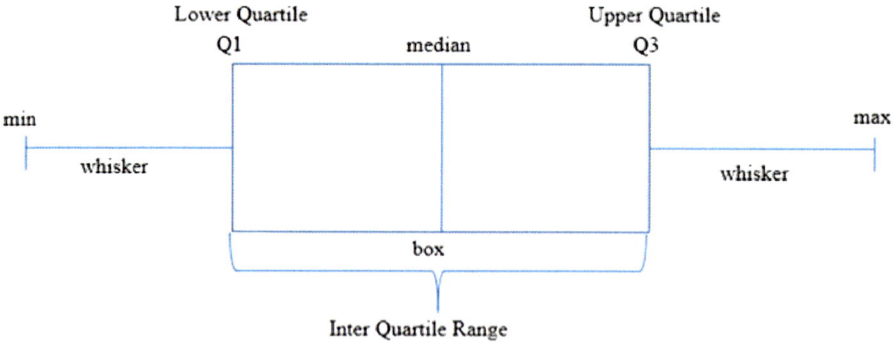

Fig. 6.4 Boxplot

Variance: $s^2 = \frac{1}{n-1}\left[\sum_{i=1}^{n} x_i^2 - \frac{1}{n}\left(\sum_{i=1}^{n} x_i\right)^2\right]$ and $\sigma^2 = \frac{1}{N}\sum_{i=1}^{n}(x_i - \mu)^2 = \frac{1}{N}\sum_{i=1}^{n} x_i^2 - \mu^2$.

Standard deviation: $s = \sqrt{\frac{1}{n-1}\left[\sum_{i=1}^{n} x_i^2 - \frac{1}{n}\left(\sum_{i=1}^{n} x_i\right)^2\right]}$ and $\sigma = \sqrt{\frac{1}{N}\sum_{i=1}^{n} x_i^2 - \mu^2}$.

Hypothesis Testing Testing is essential in all domains of research. Karl Pearson, Fisher, Neyman and Egon Pearson established hypothesis testing. It is a statistical approach for generating statistical decisions based on experimental data. It is essentially a notion about the population parameter that we make. Some of the **important terms involved in hypothesis testing** are defined below [2].

- A statistical hypothesis that believes the observation is attributable to chance is known as the **null hypothesis.** H_0**:** $\mu_1 =$**,** denotes the null hypothesis, which asserts that there is no disparity between the two population means.
- **Alternative hypotheses**, on the other hand, demonstrate that the observations are the product of a fact.
- The degree of significance with which we accept or deny the null hypothesis is described as the **level of significance.** Because it is impossible to acknowledge or deny a hypothesis with 100% correctness, we choose a degree of significance that is mostly 5%.
- A **type I error** happens when the null hypothesis is rejected regardless of the reality that the hypothesis was accurate. The letter alpha (α) represents a type I error.
- **Type II errors** occur when the null hypothesis is acknowledged even though it is untrue. Beta (β) represents type II errors.
- A **one-tailed test** happens when the produced statistical hypothesis holds only one value.
- A **two-tailed test** is when the listed statistical hypothesis undertakes a less than or greater than value.

Once the data is collected, we need to preprocess the data to assure its quality. Following are a few **measures of data quality**.

- **Accuracy** indicates to the appropriateness of the data.
- **Completeness** implies the totality of the data. Here, we check if all the values are recorded (availability).
- **Consistency** refers to the check on the collected data for any modified or dangling values.
- **Timeliness** refers to the check performed on the data for any timely updates to be done to make the data relevant and reliable.
- **Believability** refers to the trustworthiness and dependability of the data.
- **Interpretability** refers to degree of ease in understanding the data.

The **key tasks** involved in data preprocessing [3] are as follows:

Data cleaning indicates to the activity of filling in missing values, smoothening of noisy data, identification and removal of outliers, and resolving inconsistencies. Data in general is dirty. 'Dirty' data includes loads of potentially erroneous data such as instrument defective, individual or computer inaccuracy, and communication fault. Some categories of the 'dirty' data includes the following namely:

- **Incomplete** data that involves data that has a shortage in attribute values or missing attributes of importance and data containing cumulative/aggregate data. (for example, work = ' ' → incomplete data). Data in real world is not available all the time. Many tuples are left incomplete/blank with no recorded values for several attributes. Missing data can also occur owing to equipment/software failure, discrepancy found with other data that has been recorded and therefore removed, data not recorded owing to absence of clarity or misunderstanding, level of importance not determined at the time of entry and hence ignored, changed data that was not registered and so on. Under such instances, missing data must be inferred and treated accordingly.
- **Noisy data** is data that contains noise, errors, or outliers. (for example, salary = '-20' is an error). Handling noisy data can be done using one of the following methodologies such as binning (includes sorting the data and partitioning them into bin of equal frequency followed by smoothening using smooth by bin means/median/boundaries), regression (forming regression functions and smoothening by placing the data into functions), clustering (recognizing and eliminating outliers), and joint human and computer examination (identifying distrustful values with human supervision).
- **Inconsistent data** refers to data that contains discrepancies in data or values (for example, age = '42', birthday = '05/08/2010')
- **Intentional data** includes data that is disguised/default missing data (for example, filling in Feb 28 as everyone's birthday)

Data Integration implies to the task of combining multiple databases, data cubes and files.

Data Reduction refers to the action of removing all unnecessary data. Data can be reduced dimensionally, numerically and by performing data compression.

- **Dimensionality reduction** is a method for obtaining a diminished or compacted depiction of the initial data by performing feature selection (deleting features that are irrelevant or redundant) or feature extraction (translating data into features that can be used in modeling).
- **Numerosity reduction** is a data reduction strategy that uses appropriate kinds of data representation to reduce data volume. These methods might be either **parametric or non-parametric**. For parametric methods, a prototype is employed to estimate the data, thus merely the data parameters, rather than the actual data, are often retained. Histograms, clustering and sampling are non-parametric approaches for storing reduced representations of data. The models for parametric reduction are constructed using **regression and log-linear methods**. **Regression**

in turn can be as simple as linear regression (regression model with a single independent variable) or multiple regression (regression model with multiple independent variables). In a linear regression model, the data is modeled to suit a straight line, by employing the least-square approach. In a multiple regression model, the response variable '*Y*' is modeled as a linear function of a multidimensional feature vector. Based on a smaller subset of dimensional combinations, a **log-linear model** could be applied to foresee the likelihood of every data point within a multidimensional space intended for a compilation of discretized attributes.

Regression analysis is a collective term for modeling and analyzing numerical data that includes numbers of dependent variables (also called response variables or measurement) and one or more independent variables (explanatory variables or predictors). Parameters are estimated to provide better equivalent data. Usually, the best fit is tested using the method of least squares. In Fig. 6.5, the graph shows a linear regression of the form $y = x+1$. Consider the first entry, where the observed value is much lower than the expected value. A contrast among the examined value and the anticipated value (also called the residual) (observed value—expected value)) is negative. Whereas, when considering the second last entry, the observed value is higher than the expected value, and hence the difference (observed value—expected value) is positive. A sum of the squares of the residuals will help us predict the result. If the value is much lesser, then it can be inferred that the line $y = x+1$ is a best fit which means that the observed values are much closer to the expected values (minimal error).

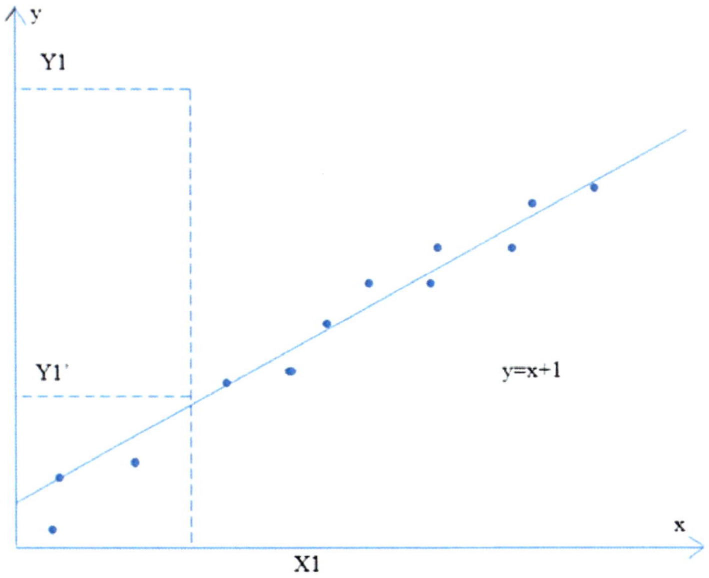

Fig. 6.5 Linear regression

- **Data compression** mechanisms help in reducing the size of files using various encoding mechanisms. They could be lossless compressions or lossy compressions. Lossless data compression use techniques to reconstruct the exact original data from compressed data thereby enabling easy and minimal data size reduction. Lossy compressions differ from the original data, but they are still useful for retrieving information. Their meanings are still comparable to the original data despite loss.

Data transformation and discretization refers to the process undertaken to change the data into forms acceptable for the mining process. This is done by **smoothing**; by building additional attributes, supplied collection of attributes known as **attribute selection; aggregation;** scaling values in a specific range (for example 6. −1.0–1.0) called as **normalization;** and **discretization** which is substituting raw numeric attribute values along with periodic or conceptual levels by the approach of **concept hierarchy generation** (transforming attributes from a lower level to a higher level in the hierarchy). For example, the characteristic 'city' can be changed to 'country.' The techniques are elaborated below:

Normalization can be done using one of the following methods:

Min-max normalization: The formula for min-max normalization is given below,

$$\text{to}\left[\underset{A}{\text{new\_min}}, \underset{A}{\text{new\_max}}\right]$$

$$v' = \frac{v - \min_A}{\max_A - \min_A}\left(\underset{A}{\text{new\_max}} - \underset{A}{\text{new\_min}}\right) + \underset{A}{\text{new\_min}}$$

For example, let the range of income be \$14,000 to \$96,000 normalized to [0.0, 1.0]. Then \$72,000 is mapped to $\frac{72{,}600 - 14{,}000}{96{,}000 - 14{,}000}(1.0 - 0) + 0 = 0.715$

Z-score normalization: The formula for z-score normalization is given below,

$v' = \frac{v - \mu_A}{\sigma_A}$,, where ($\mu$: mean, σ: standard deviation).

For example, let $\mu = 52{,}000$, $\sigma = 14{,}000$. Then $\frac{73{,}600 - 52{,}000}{14{,}000} = 1.542$.

Normalization by decimal scaling: The formula for normalization by means of decimal scaling is given as,

$v' = \frac{v}{10^j}$ where j is the lowest integer such that Max ($|v'|$) < 1.

Data Discretization can be achieved by separating the series of a continuous attribute into intervals following which discretization is done by replacing the actual data values with interval labels, splitting (top-down approach), merging (bottom-up approach), classification and recursive discretization on an attribute. There are many methods to discretize the data such as binning (unsupervised, top-down split), histogram analysis (top-down split, unsupervised), cluster analysis (unsupervised, top-down split, or bottom-up merge), decision-tree analysis (supervised, top-down split) and correlation (e.g., χ^2 (Chi-Square) test) analysis (unsupervised, bottom-up merge) of which we will elaborate on two of the methods here.

In the **binning** method [4], the actual data values are subdivided into smaller intervals known as bins, and the standard value generated for that bin is used to substitute them. This makes the input data smooth and may reduce the chances of overfilling in the case of limited data sets. There are two types of binning which are:

Binning by equal frequency where the number of items/values in every bin is the same. For example, for a data set [5, 10, 11, 13, 15, 35, 50, 55, 72, 92, 204, 215], suppose we require 3 bins then binning by equal frequency will result in 3 bins as

Bin 1: [5, 10, 11, 13],
Bin 2: [15, 35, 50, 55],
Bin 3: [72, 92, 204, 215].

In **Binning by equal width** the bins have the same width where the range of each bin is defined as [min + w], [min + 2w] .... [min + nw] where ***w*** **=** (max – min)/(no of bins). For example, for a data set [5, 10, 11, 13, 15, 35, 50, 55, 72, 92, 204, 215], suppose we require 3 bins then binning by equal width will result in 3 bins as

Bin 1: [5, 10, 11, 13, 15, 35, 50, 55, 72], where [min + w] = [5+70] = 75, and hence all numbers $\leq$ 75 fall into the first bin.

Bin 2: [92], where [min + w] = [92 + 70] = 162, hence all numbers from 92 $\leq$ 162 fall into the second bin.

Bin 3: [204, 215], where [min + w] = [204 + 70] = 274, hence all numbers from 204 $\leq$ 274 fall into the third bin. For all the above cases, we have w= [215 – 5]/3 = 70

Following the binning activity, we can proceed toward smoothening the values in the bins by following the smoothening by bin means method or bin boundaries method.

In **smoothening by bin means** method, the average of all the values in each bin is computed and all the values are replaced by the mean value.

For example, **Bin 1**: [5, 8, 9, 14],
Bin 2: [22, 21, 24, 23],
Bin 3: [25, 28, 30, 34].
On applying bin means method, we get,
Bin 1: [9], since (4+8+9+15)/4=9
Bin 2: [23, 23, 23, 23], since (21+21+24+25)/4=23
Bin 3: [29, 29, 29, 29], since (26+28+29+34)/4=29

In **smoothening by bin boundaries**, the least and maximum values are noted. The difference/distance of each value from that of the min or max (boundary values) is calculated. Whichever boundary value (min/max) is closer (less difference) to the value of interest is then used for replacement. For example, **Bin 1**: [5, 8, 9, 14], here min = 5 max = 14

Bin 2: [22, 21, 24, 23], here min = 21 max=24
Bin 3: [25, 28, 30, 34], here min = 25 max=34. By applying bin boundary method, we get,
Bin 1: [5, 5, 5, 14],
Bin 2: [21, 21, 24, 24],
Bin 3: [25, 25, 34, 34].

Correlation Analysis—Correlation depicts the degree of coordination between variables. For example, if there is frequent/more rainfall (variable 1) then the crop yield (variable 2) will be more. Here, both variables move in the same direction and hence are positively correlated. Another example is that if the price of a commodity increases (variable 1) then the purchase (variable 2) will decrease. Here, both variables move in the opposite direction and hence are negatively correlated. Next an example of no correlation will be a scenario where we cannot determine that if the size of shoes of a person (variable 1) is more, then the salary (variable 2) will be more. Here, the condition does not make sense and hence the variables are not correlated. Correlation does not imply causality in the above example. Correlation can be effectively applied on nominal data and the formula is given by the $\chi 2$ (Chi-Square) test, where $\chi 2 = \sum \frac{(\text{Observed}-\text{Expected})^2}{\text{Expected}}$. Here, if $\chi 2$ (Chi-Square) < 0 then the variables are negatively correlated. If $\chi 2$ (Chi-Square) > 0, then the variables are positively correlated, and when $\chi 2$ (Chi-Square) = 0, the variables are independent. The greater the $\chi 2$ value, the more the variables are associated.

Example 1

Table 6.1 represents the data collected by conducting a survey among 1500 students. Parenthesized numbers are the expected counts determined based on the data distribution in the two groups. The students were asked if he/she was interested in reading thriller stories and if they were interested in playing hockey. About 250 students have said yes to both (thriller and playing hockey), 200 like to read thriller stories but not play hockey, 50 like to play hockey but not read thriller stories and 1000 don't like to do both. All the above values are observed values where we have got the numbers by typically conducting a survey and logging the response. The expected values mentioned in brackets are obtained by applying the values in the following formula $\frac{\text{Column total(Sum(col.))} \times \text{Row total(Sum (row))}}{\text{Total sum}}$. For example, the expected value for [likes thriller & plays hockey] = $\frac{300 \times 450}{1500} = 90$.

Table 6.1 Survey data

	Play Hockey	Not Hockey	Sum (row)
Likes thriller	200(90)	250(360)	450
Not likes thriller	100(210)	950(840)	1050
Sum (col.)	300	1200	1500

χ^2 **(Chi-Square)** calculation is obtained by substituting the values in the χ^2 **(Chi-Square)** formula as,
$\chi^2 = \frac{(200-90)^2}{90} + \frac{(100-210)^2}{210} + \frac{(250-360)^2}{360} + \frac{(950-840)^2}{840} = 240.07$. Here, the χ^2 (Chi-Square) value is > 0 which suggests that the variables likes_thriller and plays_hockey are positively correlated.

Co-Variance evaluates the joint variability of two random variables. That is, it determines the extent to which the variables change together. However, the co-variance does not determine the dependency between the two variables.

$Cov(A, B) = E((A - \overline{A})(B - \overline{B})) = \frac{\sum_{i=1}^{n}(a_i - \overline{A})(b_i - \overline{B})}{n}$ which can be simplified as

$$Cov(A, B) = E(A.B) - \overline{A}\overline{B}$$

Example

Suppose two shares *A* and *B* have the values in one week as pairs such as: (3,5), (3,7), (4,10), (4,10) and (7,14). If the two shares are impacted by the same industry developments, is it possible that their prices also increase or go down simultaneously?

$E(A) = (3 + 3 + 4 + 4 + 7)/5 = 4.2$

$E(B) = (5 + 7 + 10 + 10 + 14)/5 = 9.2$

Cov $(A,B) = (3 \times 5 + 3 \times 7 + 4 \times 10 + 4 \times 10 + 7 \times 14)/5—4.2 \times 9.2 = 5.54$.

Since Cov $(A,B) > 0$ *A* and *B* rise together.

Classification [3] is a two-step process. After collecting the data and pre-processing it, the next step is to make a classification in the database that includes **model construction and model usage**. **Model construction** involves defining a set of pre-determined classes. In model construction, each tuple is considered to belong to the previously defined category, as determined by the class label attribute. The set of tuples used to build the model is called the training set. The model is represented as rules of classification, decision trees, or mathematical formulas.

Model usage refers to the function of distinguishing future or unknown objects based on model accuracy. The accuracy of the model is quantified by comparing the known test sample label with a distinct result from the model. An independent test set is a training set. An acceptable accuracy makes the model eligible to be used in the classification process for new incoming data. A validation set is a test set that is used to select models. Classification is a subset of supervised machine learning where the algorithms are used to perform classification and learn from a training database (such as a teacher) where the answers are already known. The algorithm then continually makes predictions which are adjusted at each iteration based on known responses/labels (training set). When the algorithm reaches an acceptable level of performance, the learning process comes to an end. There are many well-known categories summarized below.

Table 6.2 Tags associated to every phrase

Text	Tag
"A good game"	Sports
"We movie is over"	Not sports
"Very clear match"	Sports
"A memorable game"	Sports

- **Logistic Regression** forecasts a binary outcome such as Yes/No, Pass/Fail, Alive/Dead and so on. The binary effect is determined by analyzing the independent factors, which are found to fall into one of the two groups which is written as $P(Y = 1 \mid X)$ or $P(Y = 0 \mid X)$. It calculates the probability of a dependent variable Y for a given independent variable X. It can be used to ascertain the object in the picture (tree, flower, grass, etc.), and each item is given a range of 0 and 1.
- **Naive Bayes** algorithm determines whether a data falls into a predefined category or not. It can be used in text analysis to categorize words or phrases. For example, Table 6.2 shows tags—[Sports and not Sports] related to each phrase/text that can be used to classify incoming data.

  To determine whether a sentence should be labeled as "sports", we use the formula,

$$P((A|B)) = \frac{P((B|A) * P(A))}{P(B)}$$

where $P(A|B)$ is probability of A given B is true
$P(B|A)$ is probability of B given A is true
$P(A)$ is probability of A being true
$P(B)$ is probability of B being true.

- The **k-nearest neighbor** (k-NN) finds the k- nearest points for a given data point using a training data set. If $k = 1$, it is assigned to the class closest to 1. K's majority vote is used to divide it.
- **Decision tree** is a supervised learning method that can set up classes in a positive way, making it ideal for classification problems. It functions like a flow chart, dividing the data into two sections at a time, starting with 'trunk' and moving on to 'branches' and 'leaves,' where the sections are very similar.
- The **support vector machine** is a classifier that introduces a hyperplane as a reference position and classifies the data points based on the hyperplane. It aims at increasing the distance between the hyperplane and the data points.

Classifier Evaluation metrics: Once the classifier has been modeled and applied to classify the incoming samples, it is necessary to evaluate the accuracy of the classifier (performance evaluation) for which we use the confusion matrix. The Table 6.3 shows the possible outcomes for the various combinations of the actual class and predicted class. True positive (TP) represent the situation that is presented with an object

Table 6.3 Confusion matrix

Actual class\predicted class	C_1	$\neg C_1$
C_1	True Positive (TP)	False negatives (FN)
$\neg C_1$	False Positive (FP)	True Negative (TN)

Table 6.4 Example of Confusion Matrix

Actual class\predicted class	Buy_Graphics_card = yes	Buy_Graphics_card = no	Total
Buy_Graphics_card = yes	6750	50	6800
Buy_Graphics_card = no	430	2650	3080
Total	7180	2700	9880

belonging to class C_1, and the model also predicts that the object belongs to the same class C_1. False negative (FN) represents the situation that is presented with an object belonging to class C_1, and the model predicts that the object does not belongs to the class C_1 denoted as C_1. False positive (FP) represents the situation that is presented with an object belonging to class C_1, and the model also predicts that the object belongs to the same class C_1.

Table 6.4 shows an example of the confusion matrix for the variable **Buy_Graphics_card.** Here, 6750 signifies the quantity of customers who bought the graphics card and was also predicted correctly by the model (true positive). 50 denotes the quantity of customers who bought the graphics card, but the model had predicted otherwise (false negative). 430 signifies the quantity of customers who did not buy the graphics card, but the model had predicted otherwise (false Positive). 2650 represents the number of customers who did not buy the graphics card, and the model also predicted the same (True negative).

Confusion error rate, sensitivity and specificity are important aspects in determining the efficiency of a classifier which are represented in formulas given below. Table 6.5 is a condensed representation of Table 6.3 where *P* and *N* represent the total of the respective rows and *P*', *N*' and All represent the total of the respective columns.

Classifier accuracy or recognition rate is the percent of test set tuples that are accurately categorized which is calculated as,

Accuracy = (TP + TN)/All

Table 6.5 Confusion matrix (Condensed representation of Table 6.3)

A\P	*C*	– *C*	
C	TP	FN	*P*
– C	FP	TN	*N*
	P'	*N*'	All

Error rate which is 1—*accuracy,* or error rate = (FP + FN)/All

Class imbalance problem is a condition where one class may be uncommon, for example, scam, HIV-positive cases, etc. It represents a considerable majority of the negative class and a minority of the positive class.

Sensitivity is the true positive recognition rate, which is given by,

$$\text{Sensitivity} = \text{TP}/P$$

Specificity is the true negative recognition rate, which is given by,

$$\text{Specificity} = \text{TN}/N$$

Precision represents the exactness of the classification output, that is, the % of tuples that the classifier has marked as positive are positive, which is given by,

$$\text{Precision} = \frac{\text{TP}}{\text{TP} + \text{FP}}$$

Recall represents the completeness of the classification output, that is, the % of positive tuples that the classifier has labeled as positive, which is given by,

$$\text{Recall} = \frac{\text{TP}}{\text{TP} + \text{FN}}$$

From the above equations, we can deduce that there exists inverse relationship between precision and recall.

***F*-measure ($F_1 or F - score$)** is the harmonic mean of precision and recall, which is given by,

$$F = \frac{2 * \text{precision} * \text{recall}}{\text{precision} + \text{recall}}$$

F_β is the weighted measure of precision and recall. Assigning β times as much weight to recall as to precision, we get,

$$F_\beta = \frac{\left(1 + \beta^2\right) * \text{precision} * \text{recall}}{\beta^2 * \text{precision} + \text{recall}}$$

6.1.1 Data Analytics and Learning Methodologies

Both data science and data analysis interact with data, but in different ways. Data science examines data broadly, trying to determine the relationships between data attributes. Data analysis is much clearer than data science because it seeks to find relationships between specific structures. Data science can be defined as a whole, while data analysis can be defined as part of it. As mentioned earlier, the statistics are very helpful in making the analysis [5]. There are four types of data analysis, namely:

Descriptive analytics: This type of data analysis helps to understand what happened with the previous data and provides insight into past events, for example, descriptive analysis performed on data to quantify the number of deaths occurred in the state / city / s because of COVID-19.

Diagnostic analytics: This type of data analysis helps to understand the cause of the event (Why did it happen?). Data from descriptive statistics are needed to dig deeper to determine the cause of the effect, for example, diagnostic calculations performed with COVID-19 data to understand temperature, air, humidity, etc. has it contributed to the spread of COVID-19.

Predictive Analytics: This type of data analysis helps to predict future events. It uses historical data and trends to predict future outcome. For example, predictive analyzes performed on COVID-19 data lead to future spread rates and higher value depending on previous data.

Prescriptive analytics: It provides an action plan that can help increase business revenue; thus, it plays an important role in all areas of business. Today, every domain is working on understanding customer/user preferences/ways to improve services. For example, an analysis of the concept in COVID-19 can be performed to identify a vaccine and locking system that can delay the spread of COIVD-19.

Various learning techniques assist in performing data analytics which are discussed below:

Machine learning is the study of providing data and information to machines without having to organize it explicitly so that they can read and act like humans. It is one of the most sought-after methods of data analysis. There are many machine learning strategies and algorithms available, where choosing the right algorithms is important in determining the performance of the models, which requires an understanding of all the available algorithms. Machine learning algorithms are divided into three types namely, supervised learning, unsupervised learning and reinforcement learning algorithms and are used in all the data analysis categories mentioned above.

Supervised learning: As the name suggests, it works under supervision, i.e., the model predicts the classes with the assistance of a labeled database. When a model is assigned data (data with known labels and responses), it analyzes and learns to associate with labels based on factors such as size, shape, etc. Supervised learning is further categorized as **classification** and **regression**.

The problem is said to be in the form of a **classification** problem, in which the exceptions are the class, i.e., in two or more categories (for example, yes/no, true/false). Let's understand classifications with the help of an example. If we want to pre-determine if an incoming email is spam or genuine using a machine, the machine must be first taught about the various aspects in a spam email (how it looks like). Spam filters such as email content reviews, followed by a review of the email subject to check for any false information that is based on other keywords such as free, lottery, prize, etc., can be used for this purpose. Then apply filters to the restricted list globally. These are words from known spam that are already banned. All these filters create the email known as spam results. If you have low email spam points, there is a good chance that the email will reach the intended inbox of the intended subscriber. By examining the content, labels and spam result of the incoming email, the trained algorithm makes an informed decision on whether the mail should go to the spam folder or the intended recipient's inbox. The most well-known algorithms are the decision tree, the random forest, the Naïve Bayes, *k*-nearest neighbor and the support vector machine algorithms to support all the class-based/classification algorithms.

A problem is called a **regression** problem, when a change in one variable is associated with the change in another variable, given that the two variables are related. To understand this, lets take the example of two related variables, namely temperature (independent) and humidity (dependent). The humidity is said to decreases with a rise in temperature which states that both are associated. We can feed this information to the regression model which will learn the relationship between the two variables upon appropriate training and will further be able to predict the humidity for a given temperature.

Some of the applications in which supervised learning is used are risk assessments to assess risk in financial services and the insurance domain to reduce corporate risk portfolio. It is also used in image classification where the application can detect a person's photo based on a tagged photo album. It is also used to detect fraud to determine if a user's activity is genuine and tangible.

In **unsupervised learning**, there is no surveillance, that is, no training is provided to the machine, facilitating it to operate on unencrypted data. The machine tries to detect patterns and deliver the answer. Let's take an example when we feed a machine with data (in this case, spoons and knives), but we do not tell the machine their labels (that is, we do not mean spoons or knives). The machine detects patterns from a set and classifies them based on the identified patterns, similarities and so on. Most applications use the most popular k-means algorithm for model building. Also, unsupervised learning can be further collected and consolidated.

Clustering is basically where the machine creates groups based on data behavior. Let's understand clustering with the help of an example. To reduce its churn level, a communications company studies customer behavior based on the average duration of calls and internet usage. It has been noted that while some of the customer call times are high some have difficult internet usage. Customers recruited based on their behavior are recognized, and a strategy is devised to reduce the level of churn and increase profits. Group A customers are those who use additional internet data and have a longer call time. Group B customers are those who utilize the Internet a lot, and Group C customers have the highest call time. Thus, as a strategy, Group B customers can be presented with more data profit plans, while Group C customers can be presented with inexpensive call rates to purchase their reliability.

Association is a machine learning phenomenon which is rule-based. It finds its application in larger databases wherein it helps in spotting interesting relationships between variables. For example, identifying the customers who make purchases of same products every time falls under clustering. However, identification of the products that are repurchased (or purchased often) by customer/s is association. Let's understand the connection with the example. Suppose customer 1 goes to a store and buys bread, milk, fruit and wheat. Customer 2 buys bread, milk, rice and butter. Now, when buyer 3 goes to buy bread, he is more likely to buy milk. Thus, relationships are identified based on the behavior/pattern of customers and recommendations are made based on the identified relationships.

Reinforcement learning is a machine-based machine learning method that uses a set of non-labeled data as input. In this case, the model/agent learns from its mistakes. A positive response/reward is given if the action is right, while a negative response/punishment is given if the action is wrong. The next agent action is determined by the input received. Thus, the agent improves its performance based upon the feedback (positive and negative) it receives upon interacting with the environment. It finds its application in video games and robots.

Deep learning is a subset of machine learning which is a part of artificial intelligence [6]. Artificial intelligence aims at mimicking human behavior. Machine learning uses algorithms to gain knowledge on the working environment. Deep learning is a form of machine learning that works toward the development of a human brain structure with sufficiently incorporated knowledge, also called as Artificial Neural Networks (ANN). For a system to be able to distinguish between tomatoes and cherries, if trained using machine learning, we need to feed in some information to the system based on which the categorization can be done. This could include features such as the size, shape, type of stem, etc. But, with deep learning these features will be automatically selected by the ANN without human intervention. However, it requires the availability of huge amount of data set to train the machine. There are a variety of mechanisms in deep learning which are as follows.

Artificial Neural Network (ANN): Input layer, multiple hidden layers and outer layer form ANN, which is a multi-layer network. Each node in one layer is connected to the nodes of the next layer. The weight of each layer is calculated, and each layer node calculates the total weight of all the input edges and sends it to the activation function for the next layer node. As a result, each node acts as a node input in the next layer, and the number of hidden layers can be raised to improve performance. Gradient descent with backpropagation is applied to ANN. Because the inputs and outputs of each layer may vary in size, ANN makes indirect changes. Text data is where ANN is mostly used.

Convolution Neural Network (CNN): In image data sets, CNN is used to distinguish between different images. CNN has several layers, which include convolution and integration. The convolution layer detects pre-installed images and extracts image elements such as edges, color and more. Many layers of convolution can be added to extract high-quality features, as the basic features are found in the first layer of convolution. The pooling layer is the next layer, which helps to reduce the size of the elements by reducing the size. The pooling layer can reduce the number of calculation tools needed for analysis.

Recurrent Neural Network (RNN): In a RNN, every layer is independent and does not maintain records of the previous layer's outcome. However, it overcomes the problem that exists due to the hidden layers as opposed to ANN. The output of the previous layer is used as input for the next layer. Every node in that layer remembers the output as it is allocated memory called as the hidden state. It finds its application in time series data due to the ability to remember information at each layer.

6.1.2 Data Visualization Tools and Data Modeling

Visualization is usually any way to create images, drawings, or animations to convey a message. A data scientist performs data recording to shape business data and makes digital diagrams (data) to reveal hidden patterns/details. Data viewing is of two types.

Explanatory: The purpose of explanatory visual images is to tell stories—they are carefully constructed to present important findings.

Exploratory: Exploratory visuals create connections into databases. It helps users to scan data, allowing them to access their data. They are typically interactive visualizations. Although there are many Python libraries for plotting, only a handful can create interactive charts, which we can embed online and spread.

Many Python libraries help view data [7]. In this section, we discuss Python libraries that are widely used for examples.

- **Matplotlib** is a two-dimensional plotting library. We can generate plots, bar charts, histograms, power spectra, stemplots, scatterplots, error charts, pie charts and much more with a few lines of code using the matplotlib library. Let's take a look at a simple example where matplotlib is used to display a graph using a Jupyter Notebook. The code is represented in Fig. 6.6.

The magic operation starts with the % symbol. % Matplotlib inline is a function that visualizes a graph in the Jupyter Notebook itself. Next, we import the matplotlib and display the graph as shown in Fig. 6.6.

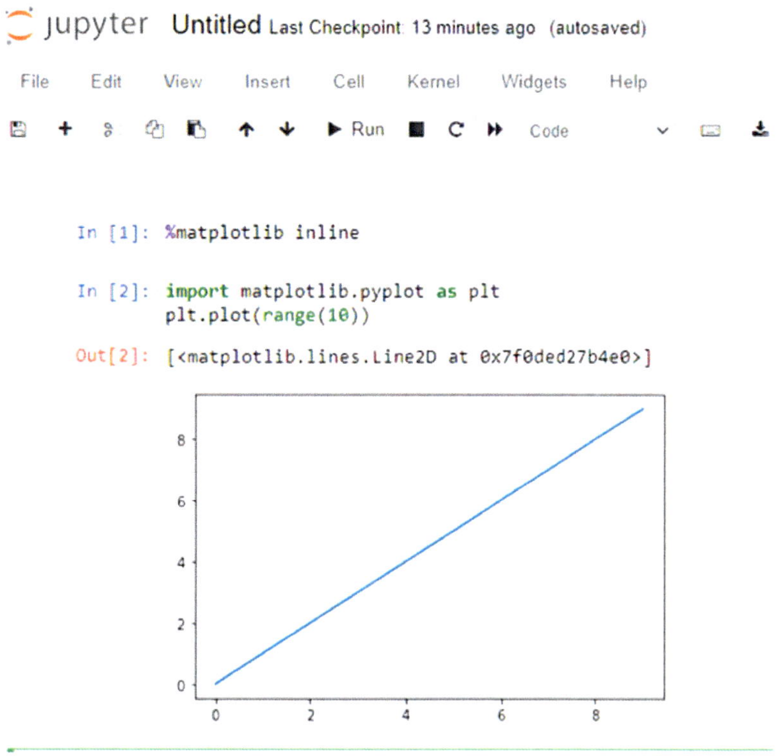

Fig. 6.6 Matplotlib function

- **Seaborn** is a matplotlib-based library. It helps to create attractive and informative math graphics in Python. Seaborn aims to make visualizations a central part of data analysis and understanding. The plotting functions apply to the data frames and perform internally the required integration and calibration of the mathematical model to produce informative plots. Figure 6.7 shows a simple example using a Seaborn library and the output is shown in Fig. 6.8.

- ***Plotly*** is a web-based toolkit to develop data visualizations. It supports visualizations such as scatterplot, error bars, line charts, bar charts, histogram, multiple axes, box plots, subplots and much more.
- To use plotly in anaconda, we first need to install plotly in anaconda. For this, first we need to run the anaconda command prompt and execute the command as given below:

  conda install -c plotly
  The Fig. 6.9 shows the execution of the command in anaconda command prompt

- We can check if the installation was successful by opening the Anaconda Navigator click on environments as shown in Fig. 6.10.

```
In [1]: import pandas as pd
        import seaborn as sns
        sns.set(color_codes=True)

        Matplotlib is building the font cache; this may take a moment.

In [17]: health = pd.read_csv('heart.csv')

In [18]: health.head()

Out[18]:
```

	age	sex	cp	trestbps	chol	fbs	restecg	thalach	exang	oldpeak	slope	ca	thal	target
0	63	1	3	145	233	1	0	150	0	2.3	0	0	1	1
1	37	1	2	130	250	0	1	187	0	3.5	0	0	2	1
2	41	0	1	130	204	0	0	172	0	1.4	2	0	2	1
3	56	1	1	120	236	0	1	178	0	0.8	2	0	2	1
4	57	0	0	120	354	0	1	163	1	0.6	2	0	2	1

Fig. 6.7 Seaborn Library

- Go to home and launch the Jupyter Notebook as shown in Fig. 6.11.

The code represented in Fig. 6.12 is a simple example showing the usage of plotly in Jupyter Notebook. The commands in the first cell display the version of plotly. The next cell is basically to set up the Jupyter Notebook. Finally, the last cell plots a simple line graph using the iplot function.

- ***Ggplot***

Ggplot is a Python application for the grammar of graphics on R programming language. Ggplot's method is not designed for creating highly customized graphics.

In this example we use the Fisher's Iris multivariate data set. There are 150 records in the collection, divided into five categories: petal length, sepal length, sepal width, petal width and species. The data set includes 50 samples from each of the three Iris species. It became a standard test case for many machine-learning statistical classification approaches, such as support vector machines. The iris data set is commonly used as a machine learning beginner's data set. The data set is presented in *R* base and Python's Scikit-learn machine learning package, so users can access it without having to look for it elsewhere. Before we start, **in** order to use ggplot we first need to install *R* in the Jupyter Notebook. The steps are as follows:

- Navigate to the Anaconda Navigator as shown in Fig. 6.13.
- Choose the environment tab—click create—name the environment as 'r_tutorial'—select Python and *R* along with the appropriate versions—click on create as shown in Fig. 6.14.
- Click on Home tab—select the created environment 'r_tutorials' from the drop-down beside 'Applications on' and launch the Jupyter Notebook as shown in Fig. 6.15.

```
In [19]: health.info()

<class 'pandas.core.frame.DataFrame'>
RangeIndex: 303 entries, 0 to 302
Data columns (total 14 columns):
 #   Column    Non-Null Count  Dtype
---  ------    --------------  -----
 0   age       303 non-null    int64
 1   sex       303 non-null    int64
 2   cp        303 non-null    int64
 3   trestbps  303 non-null    int64
 4   chol      303 non-null    int64
 5   fbs       303 non-null    int64
 6   restecg   303 non-null    int64
 7   thalach   303 non-null    int64
 8   exang     303 non-null    int64
 9   oldpeak   303 non-null    float64
 10  slope     303 non-null    int64
 11  ca        303 non-null    int64
 12  thal      303 non-null    int64
 13  target    303 non-null    int64
dtypes: float64(1), int64(13)
memory usage: 33.3 KB

In [20]: sns.barplot(health['chol'],health['oldpeak'])
Out[20]: <AxesSubplot:xlabel='chol', ylabel='oldpeak'>
```

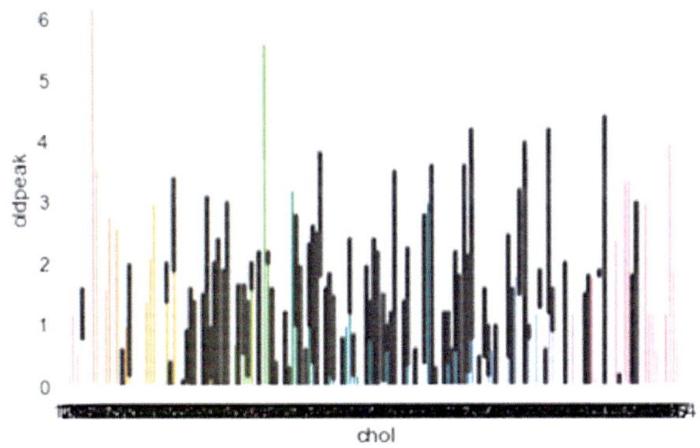

Fig. 6.8 Seaborn Library output

- We use the dplyr to read and manipulate the data set as shown in the figure. Upon clicking run, the iris data set is displayed as shown in Fig. 6.16.
- We can make a scatter plot comparing the length and width of the sepals of three different iris species using ggplot as shown in Fig. 6.17.

- ***Tableau***

Tableau is a data visualization software that helps users see and understand data in minutes [8]. It allows us to import data from various sources and use them to visualize and analyze the data and to develop better solutions and business plans. The

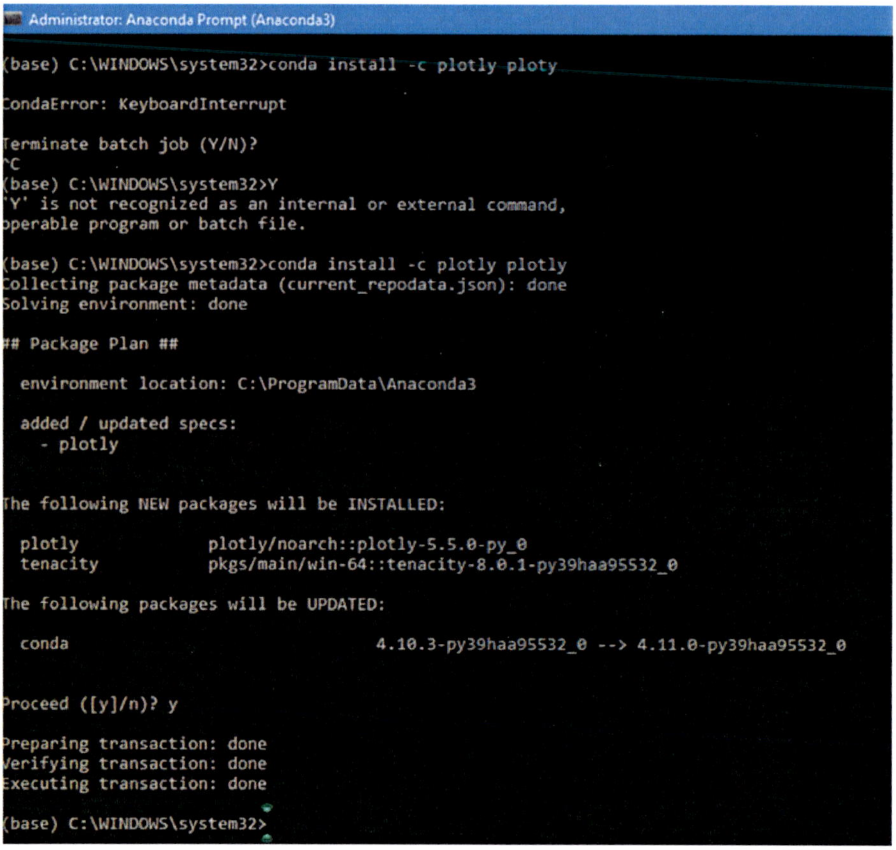

Fig. 6.9 Plotly installation using anaconda command prompt

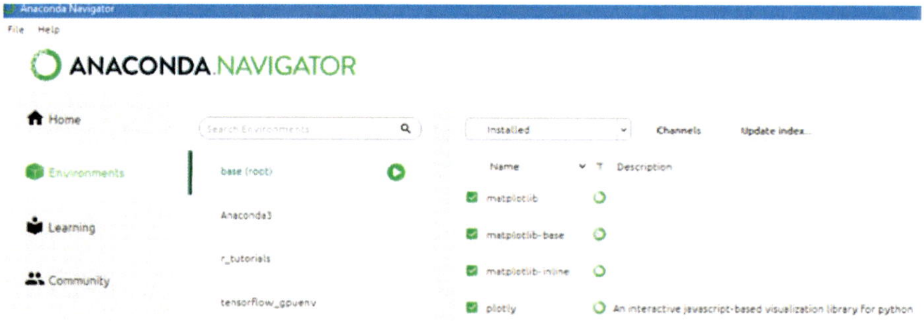

Fig. 6.10 Successful installation of Plotly shown in Anaconda navigator

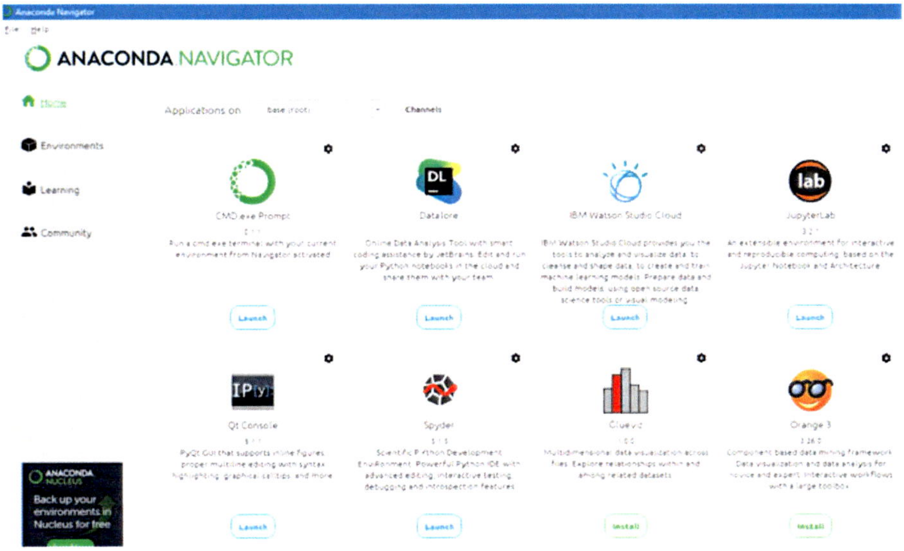

Fig. 6.11 Jupyter Notebook Launch from Anaconda navigator

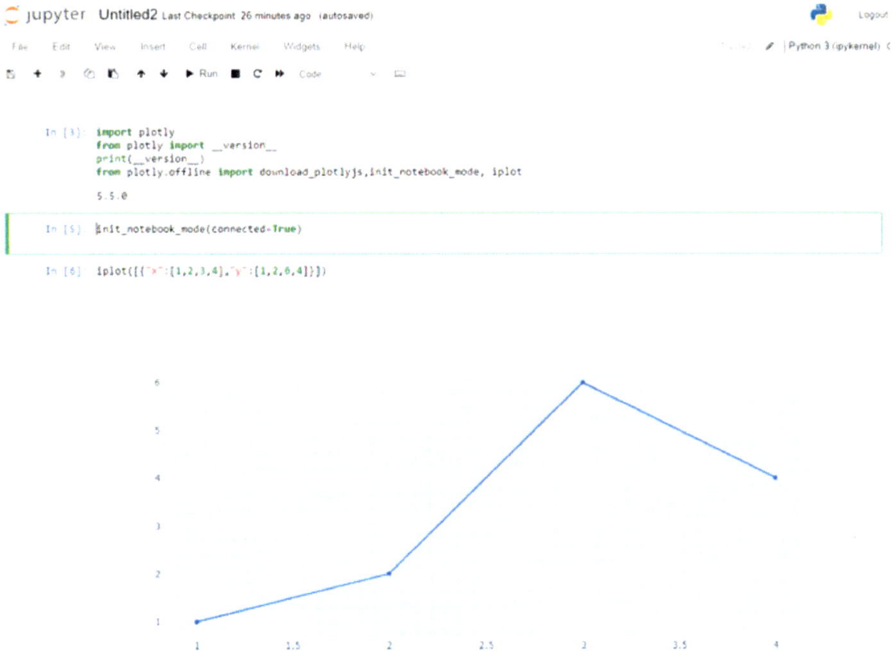

Fig. 6.12 Plotly in Jupyter Notebook

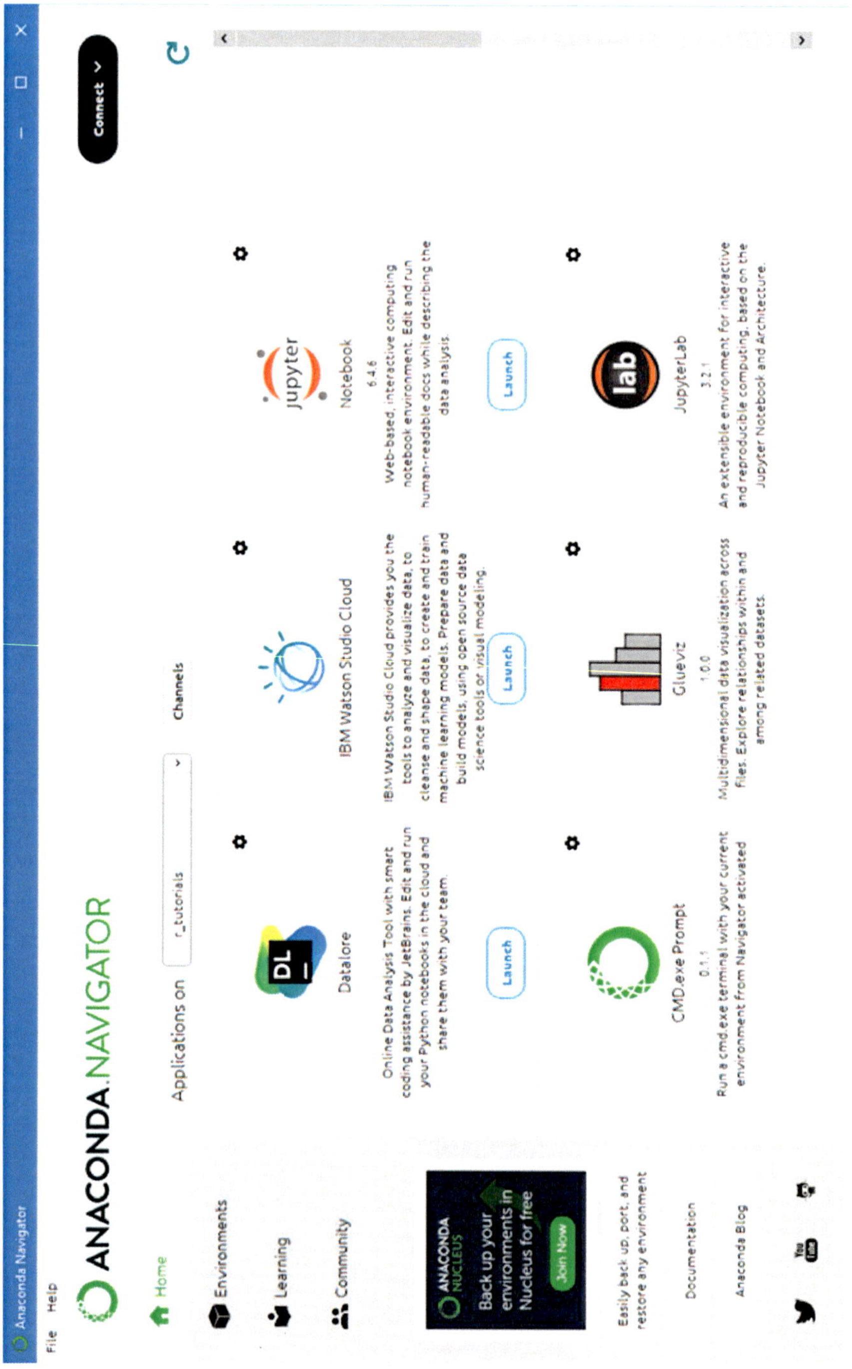

Fig. 6.13 Anaconda navigator

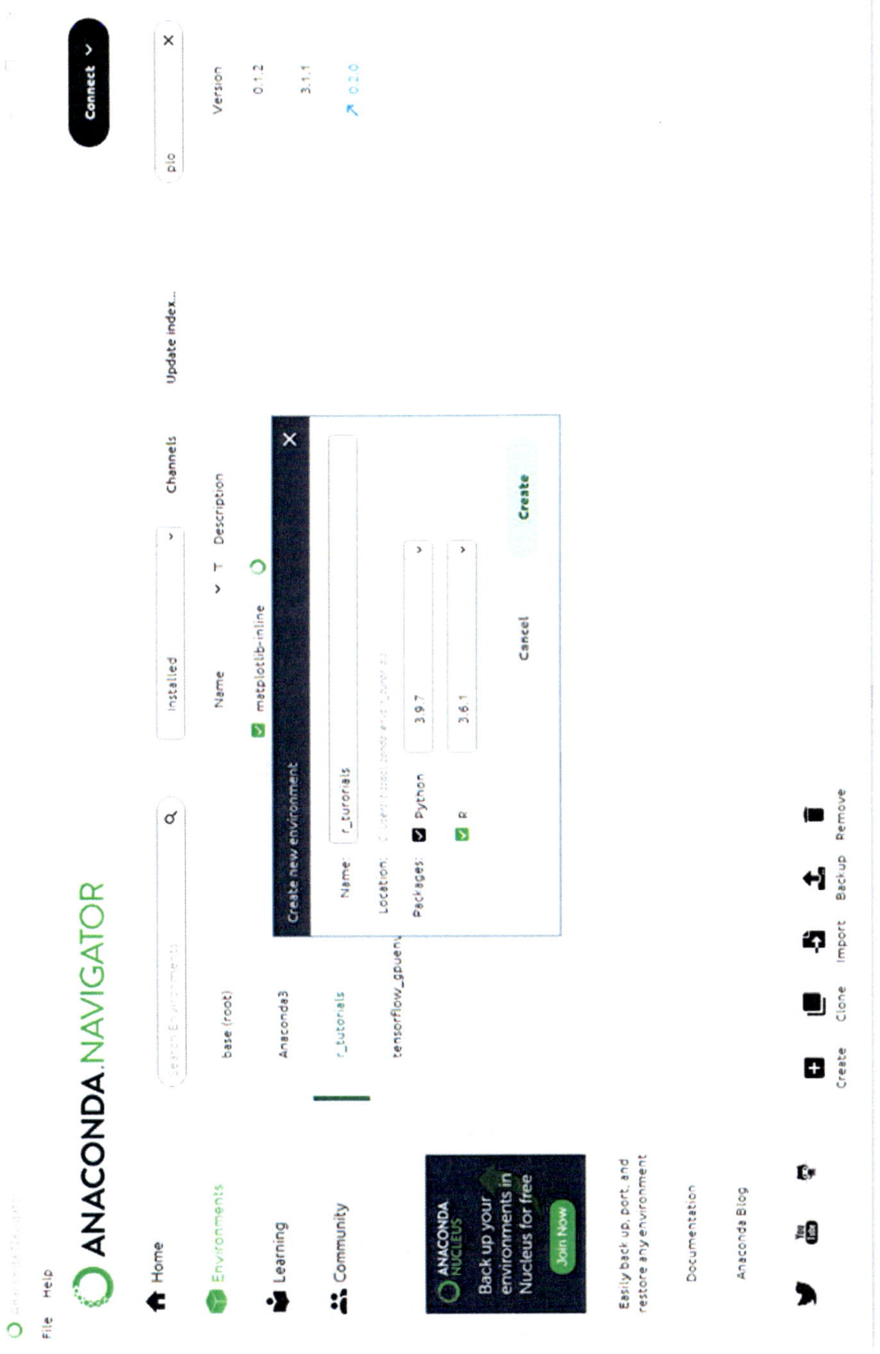

Fig. 6.14 Installing *R* in anaconda

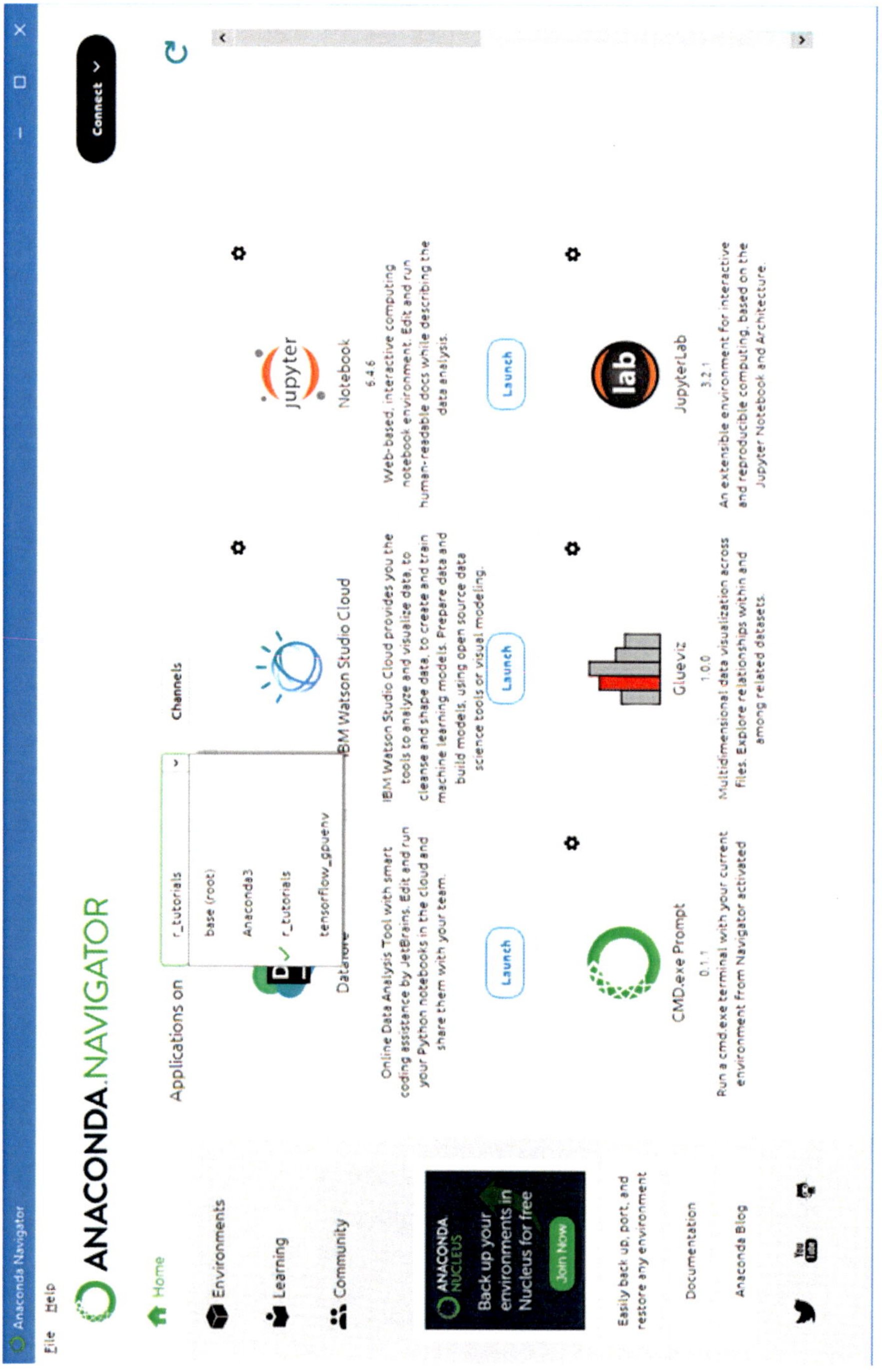

Fig. 6.15 Launching Jupyter notebook with *R*

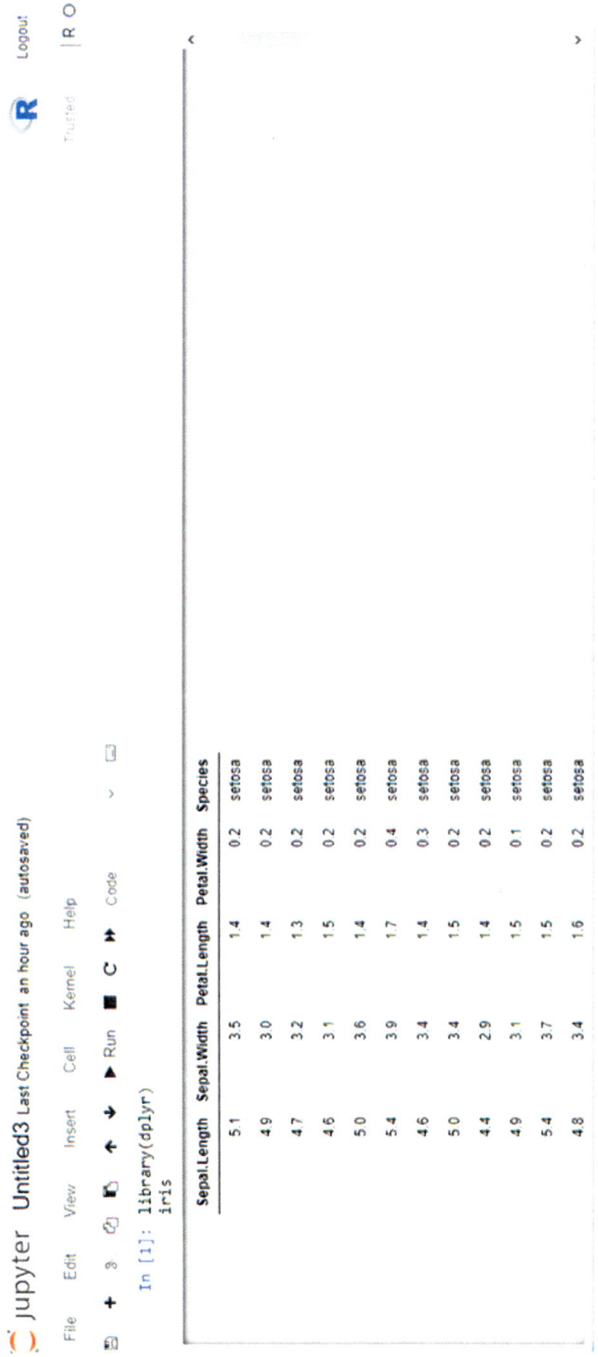

Fig. 6.16 Reading the iris data set

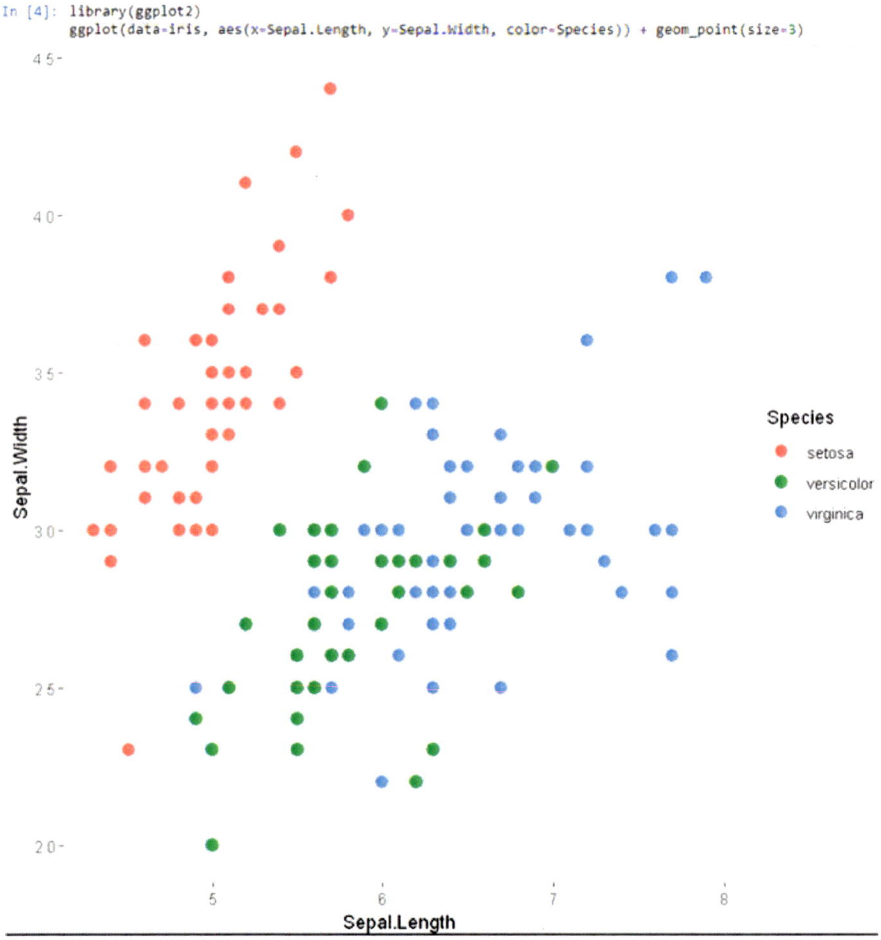

Fig. 6.17 Scatter plot for iris data set

following screenshots show the typical steps involved in using tableau software. From the variety of options provided by tableau to import data, we choose the simplest from, Microsoft Excel, for this example. Click on Microsoft Excel from the left pane—select the excel sheet of interest—click open as shown in Fig. 6.18.

Once the excel sheet is loaded, the following screen as shown in Fig. 6.19 is displayed. The left pane displays all the elements/tables that are involved in the sheet that has been loaded. The workspace to the right supports in creating models and visualizing the elements of the tables involved.

We can then drag and drop the elements from the left pane over to the workspace and create models. We can also establish joins and relations between the tables as shown in figure. These relations and joins are done in the physical layer. We then click on the sheets tab mentioned at the lower end as shown in Fig. 6.20.

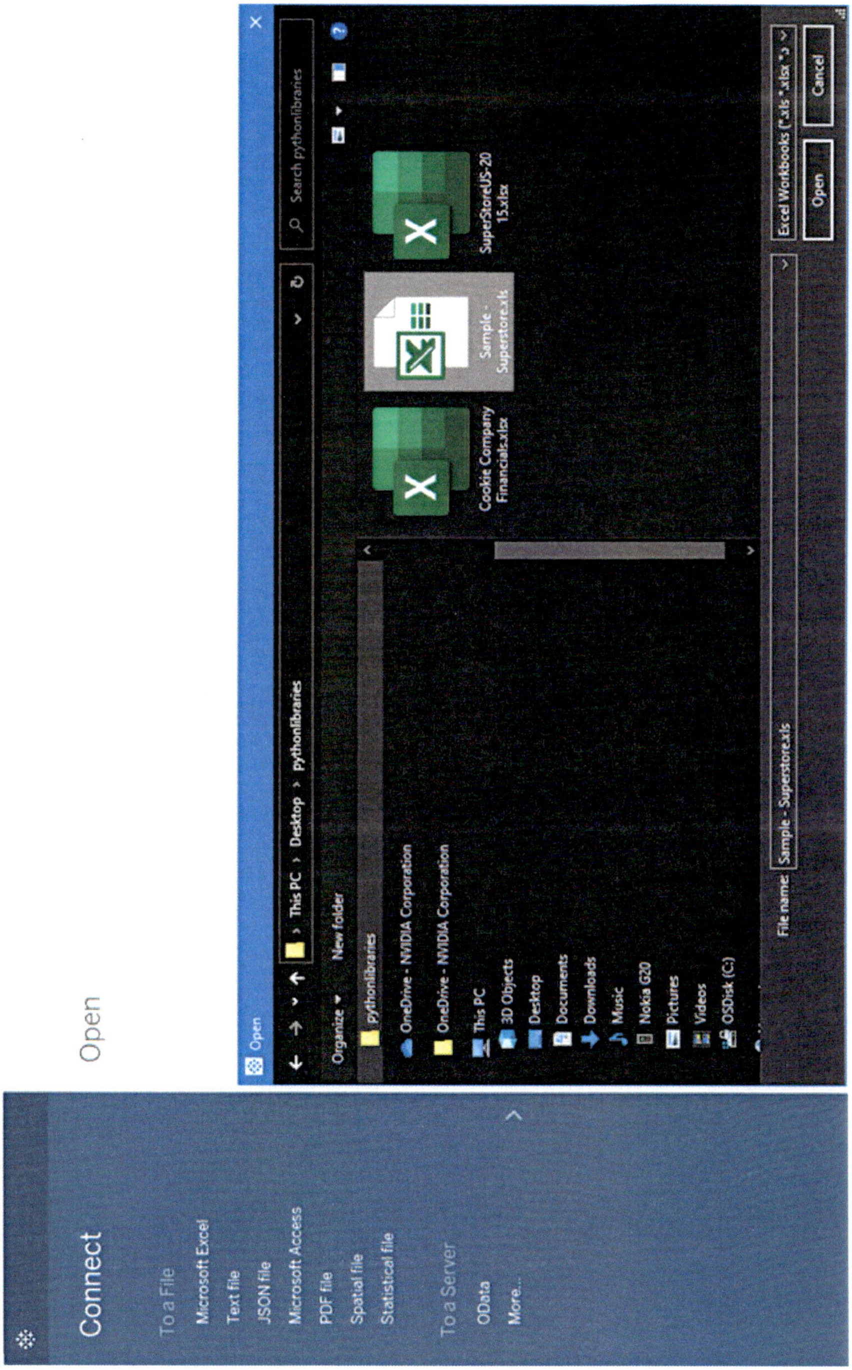

Fig. 6.18 Loading excel sheet to Tableau

Fig. 6.19 Home page/workspace after loading the file

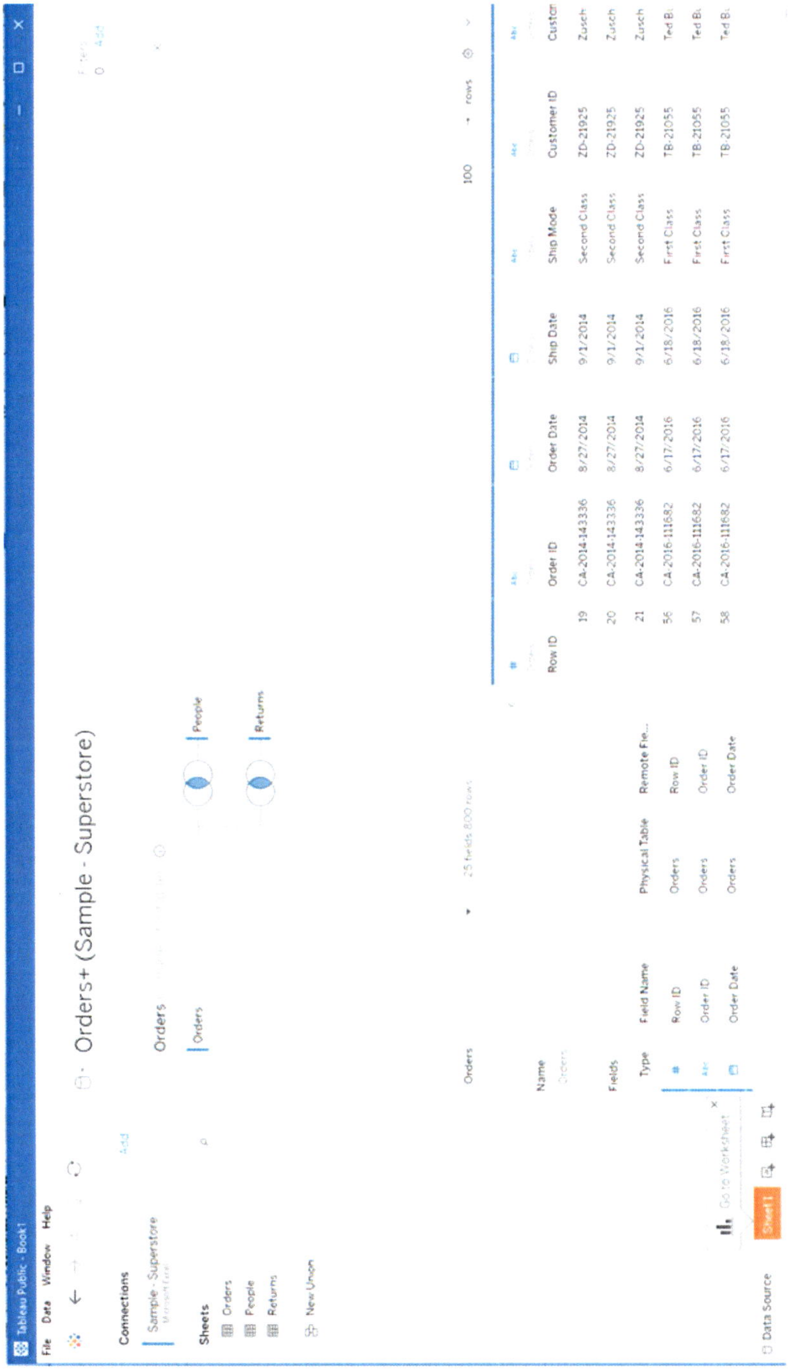

Fig. 6.20 Migrate to sheet1

In the Fig. 6.21, on the left pane we have the data elements displayed, shelves (columns, rows, page, filters and marks) which assist in developing visualizations. The show me tab displays all the possible charts that can be adopted to make attractive visualizations.

To create a bar chart that depicts the sales by category, we drag and drop the category element from the left pane into the rows shelf (top) and the sales element to the columns shelf (top). Tableau automatically populates a bar chart. We can choose other visualization forms by clicking on the icons under the show me pane on the right. We can additionally view the chart for specific regions by adding the regions element to the filters shelf. The dialog box pops up from which we can choose the regions as shown in the Fig. 6.22.

We create another sheet which displays the sales by date/month as shown in the Fig. 6.23. The order date is dragged and dropped in the columns shelf. The order data element is further expanded to show the date, quarter, etc. We remove the quarter element for better visualization. The sales element is placed in the rows shelf and a line graph is displayed as shown in the Fig. 6.23.

Further, in order to apply the filters to the second worksheet (sales_by_date) also, right click on the regions element in the filters shelf in the first sheet (sales_by_order)—select Apply to worksheets—select All using this data source —click ok as shown in Fig. 6.24. The effects will be applied across all worksheets using the same excel sheet and can be reflected in the dashboard.

To create a dashboard, we click on the dashboard icon in the lower pane beside the worksheet icon. On the left pane, the two worksheets (sales_by_category and sales_by_date) are reflected. Drag and drop the sheets into the workspace. All elements including the filter are reflected in the dashboard as shown in Fig. 6.25.

By clicking on the funnel icon beside the charts, the charts specific points can be selected to act as filters that affect the visualization of the other related charts as shown in the Fig. 6.26.

- ***Microsoft PowerBI*** is a data visualization tools that helps in interactive and informative dashboards that help in providing insights to businesses [8].To start with we launch Microsoft PowerBI and then click on the Get data icon. The dialog for Get data opens and then we can choose the source to connect to from the list that appears. Here we take up a simple example by clicking on the Excel workbook and then click on connect as shown in Fig. 6.27.

We then select the spreadsheet of our choice for visualization and load it into PowerBI as shown in Figs. 6.28 and 6.29.

At this point, we could check the required table from the left-hand side and click on load. But before that lets look on how we can transform (modify data before loading to PowerBI and build models with them) the data. Click on transform data button from the above screenshot and perform modifications as required. With the Cookie company data, we change the type of units sold data from decimal to whole numbers (right click on the units sold column—change type—whole numbers). The

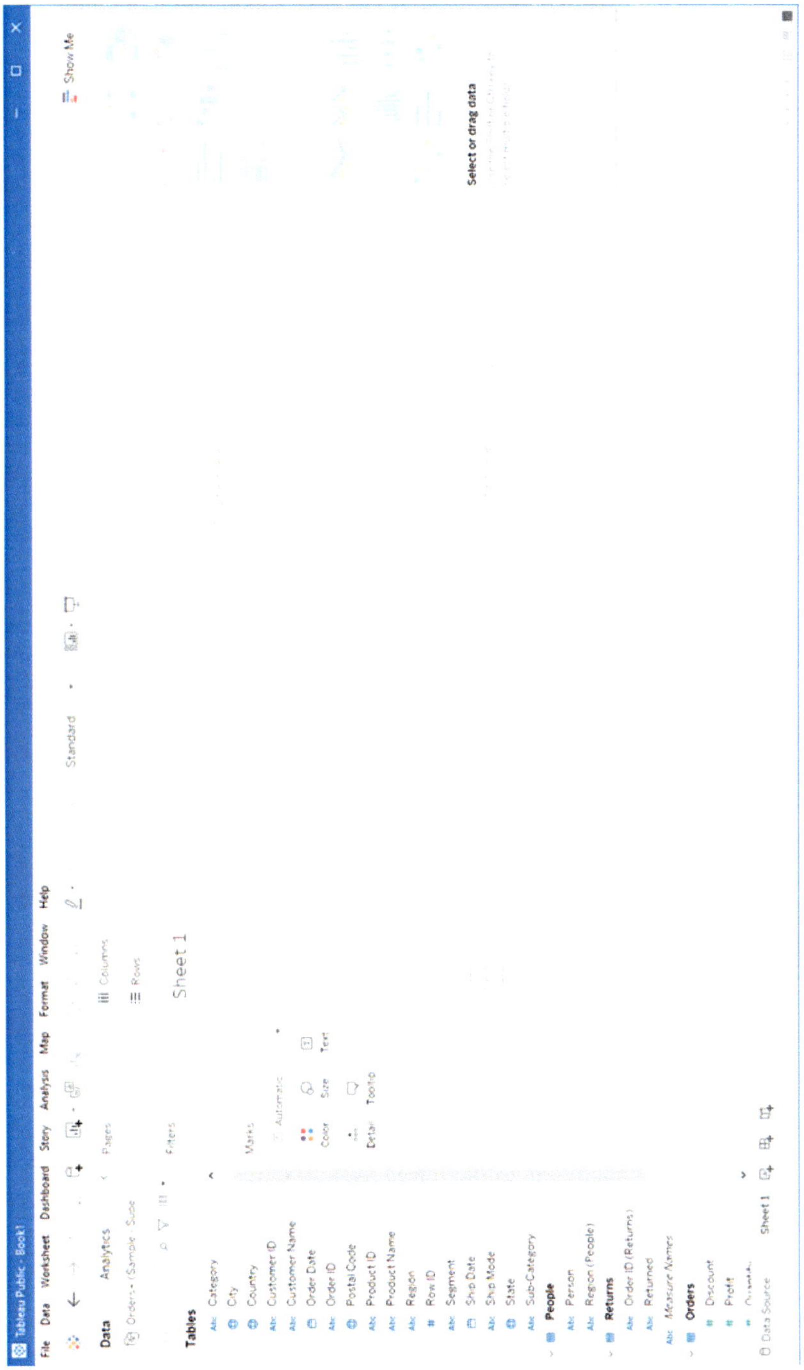

Fig. 6.21 Components of the workspace (sheet)

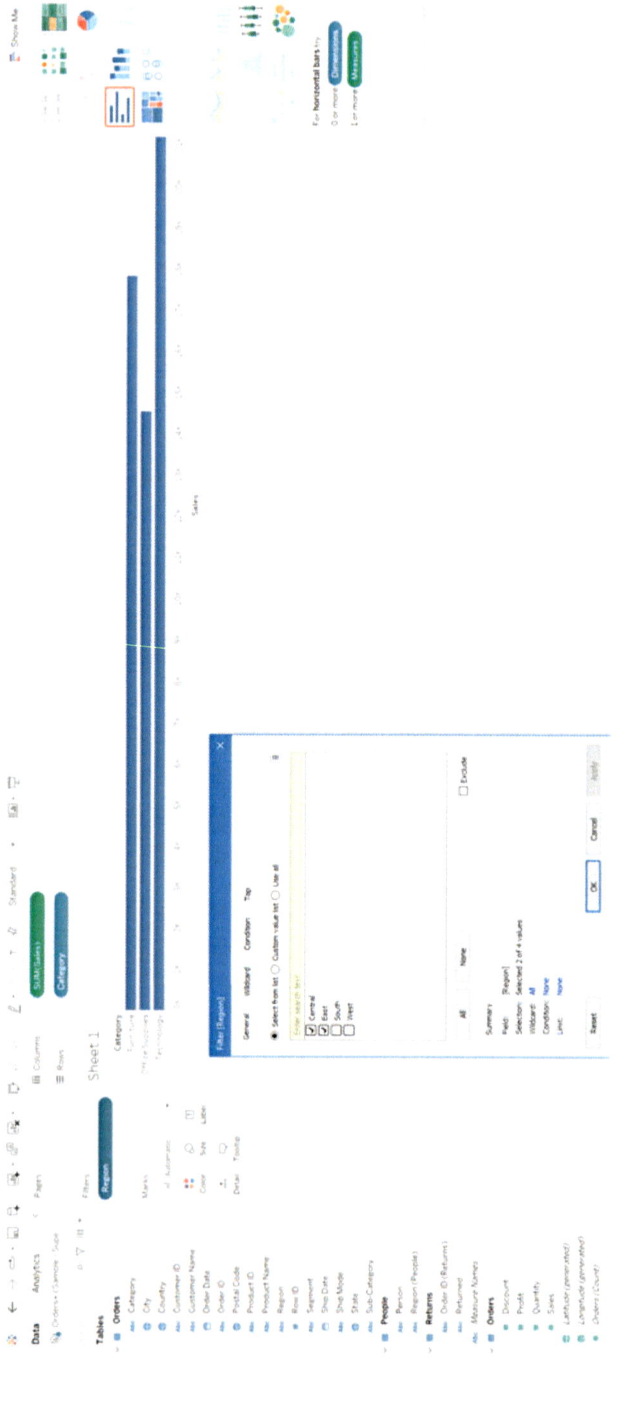

Fig. 6.22 Bar chart creation

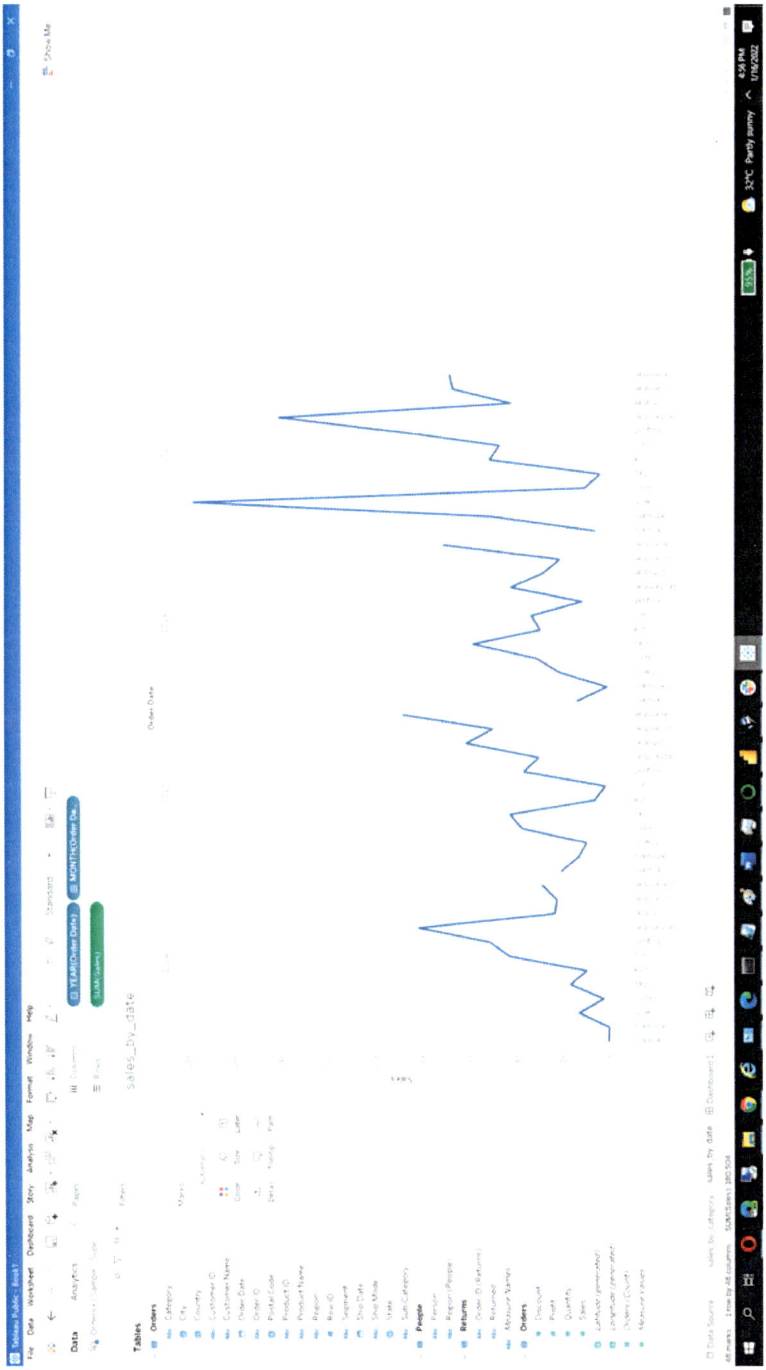

Fig. 6.23 Line graph for sales by month

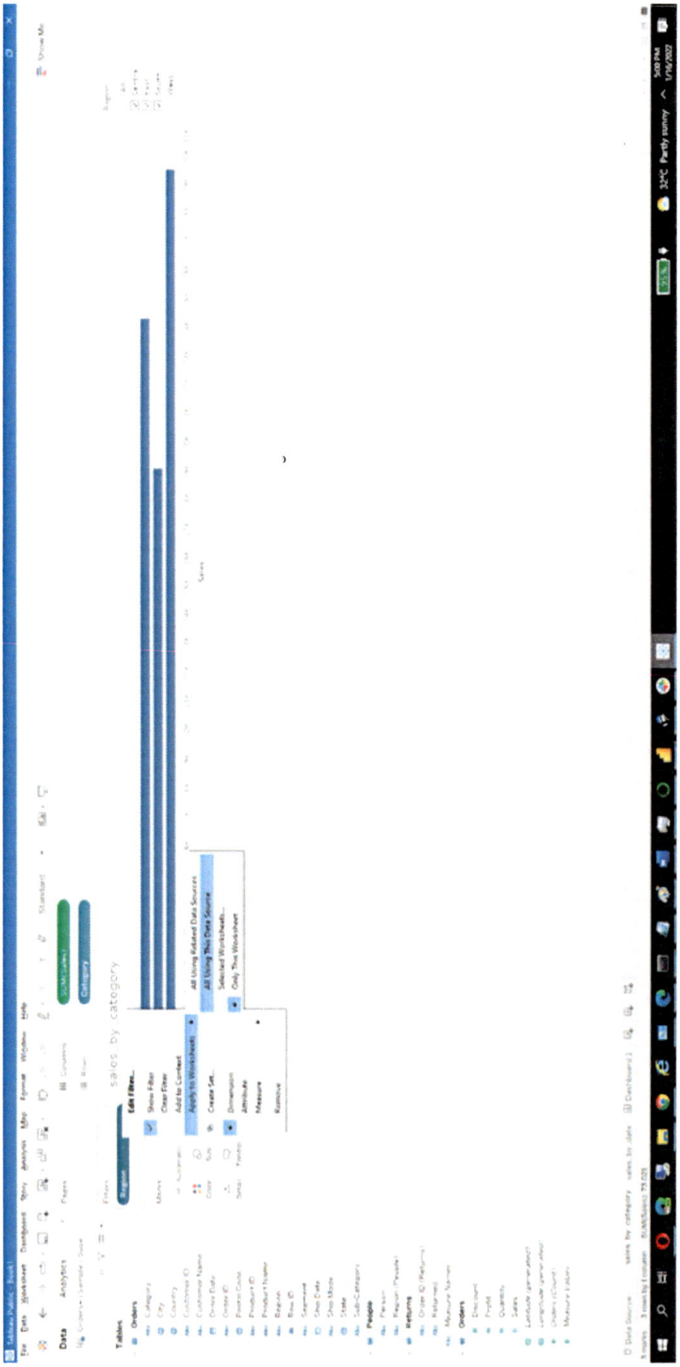

Fig. 6.24 Filter application

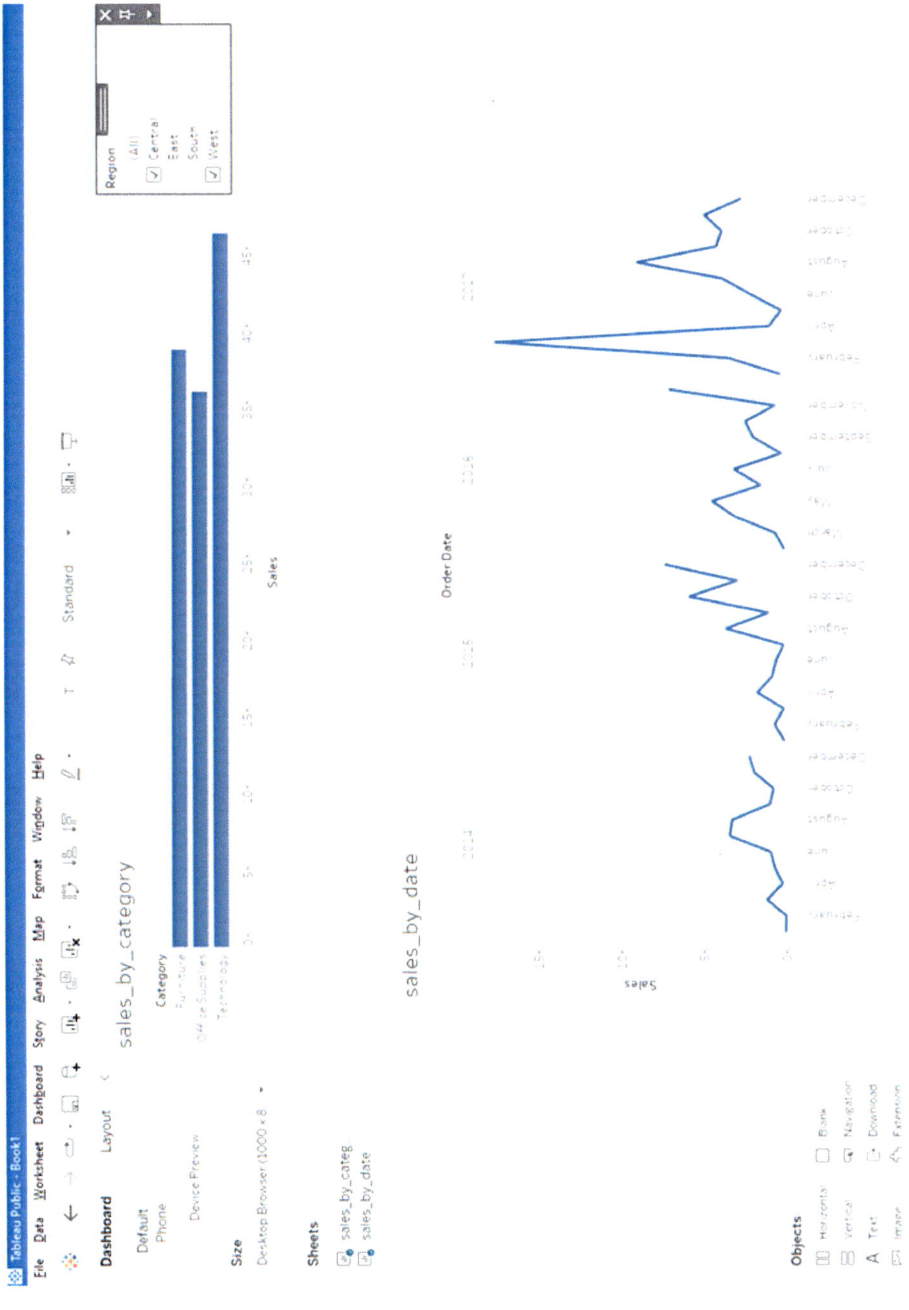

Fig. 6.25 Dashboard creation

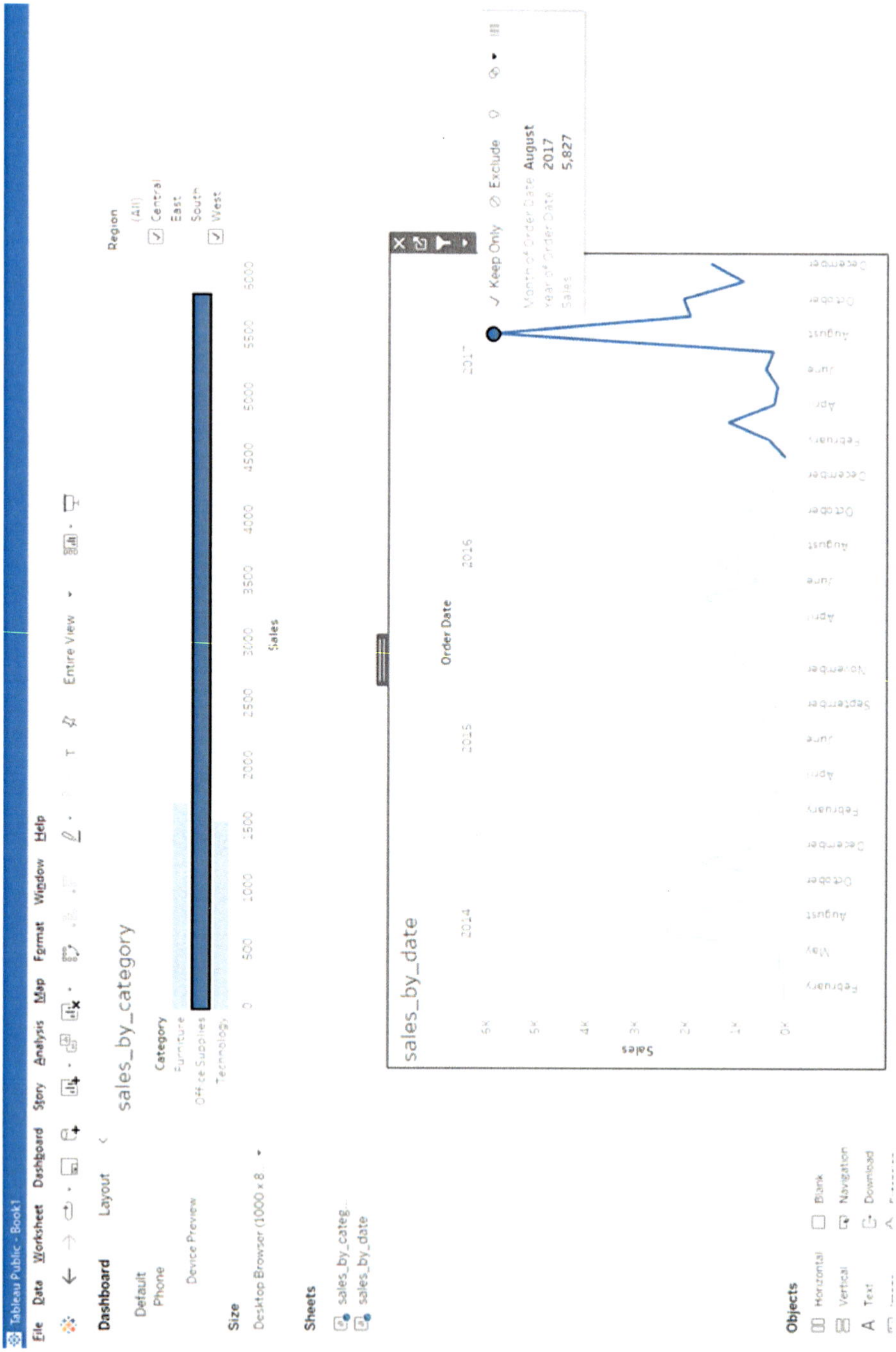

Fig. 6.26 Filter application using specific points on the chart

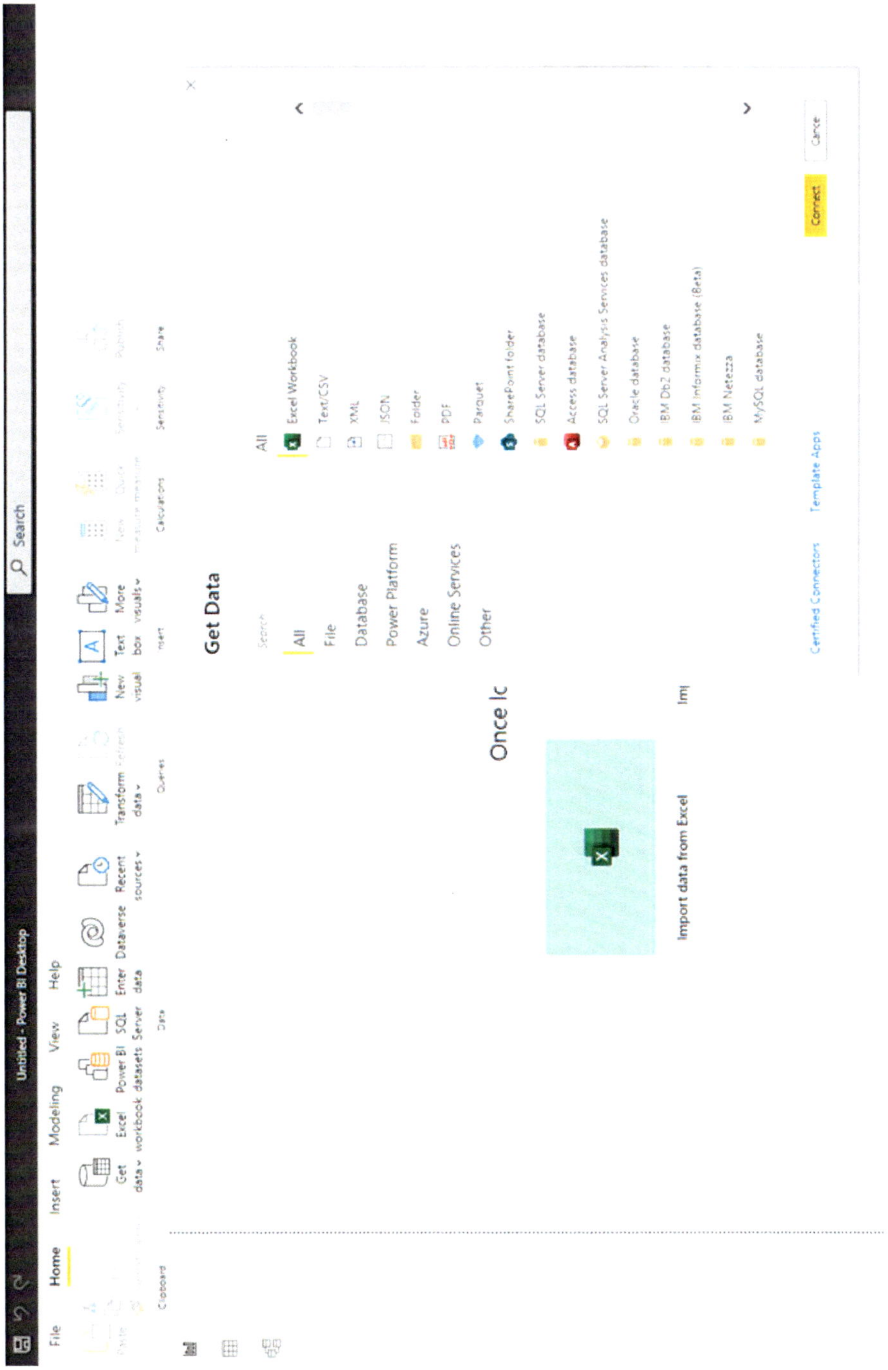

Fig. 6.27 Launching powerBI

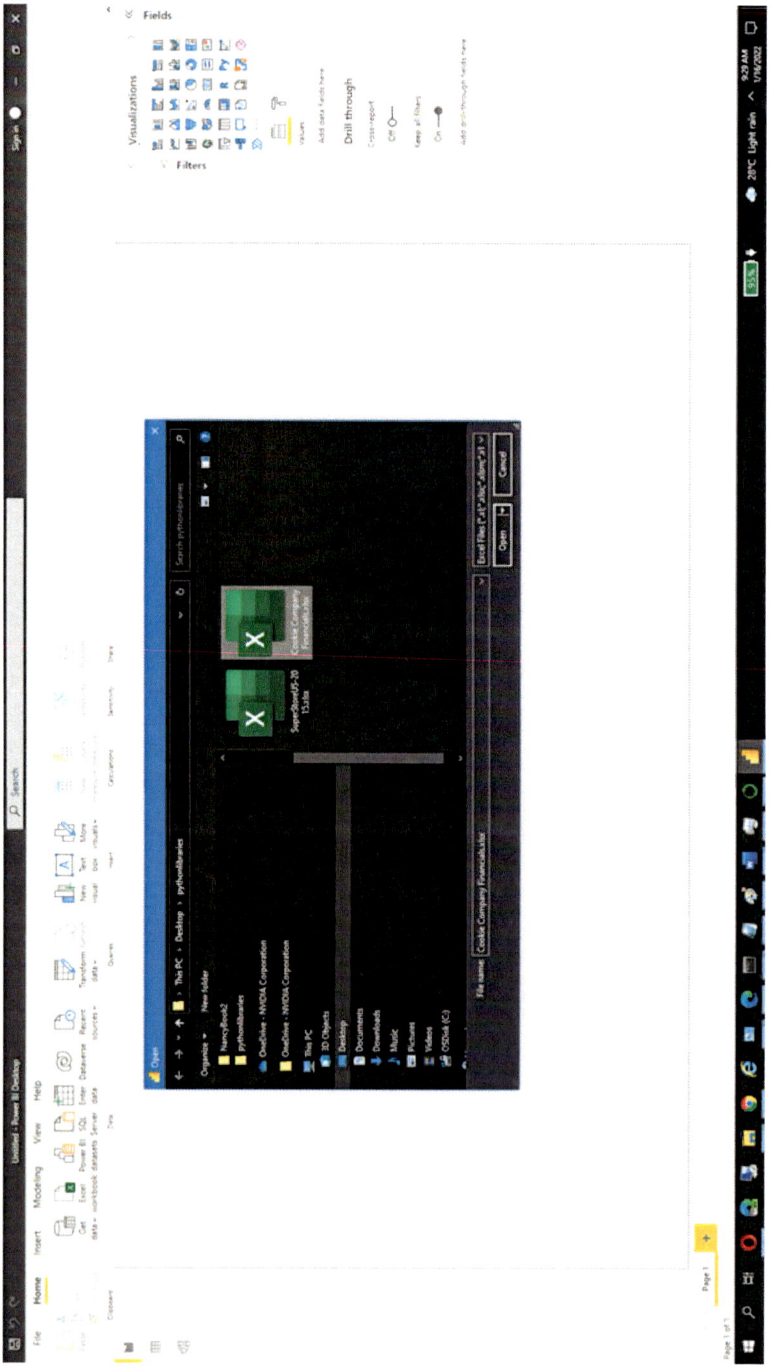

Fig. 6.28 Loading data

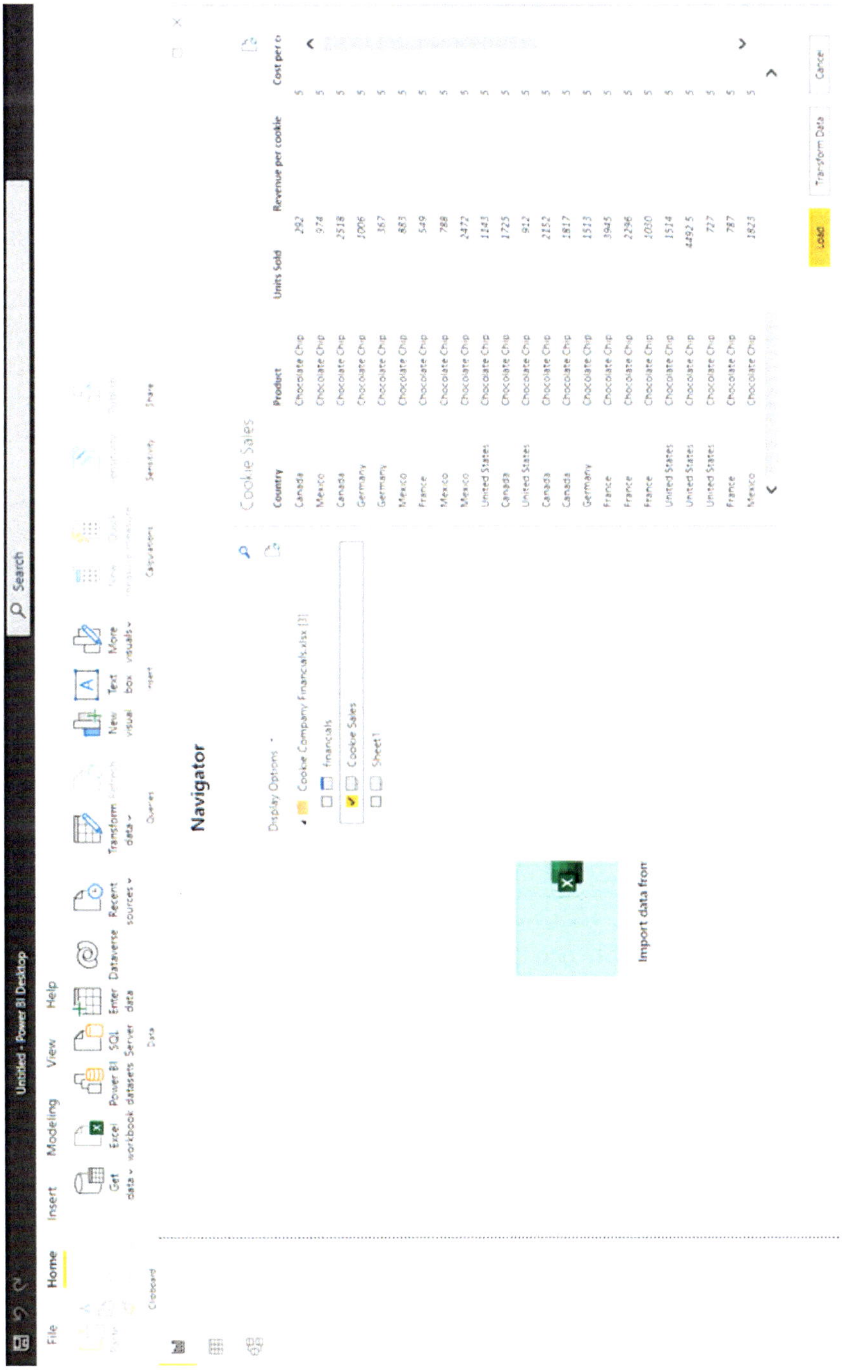

Fig. 6.29 Loading data

transformations that we have made can be seen on the right pane under the applied changes section as shown in Fig. 6.30.

Next before loading, we can also include data from the web as shown in Fig. 6.31. Here, we would like to append the population of each country where the Cookie company has its branches from an existing Wikipedia page. To do this, we click on new source drop down—select web—enter the web page URL in the dialog box that pops up—ok—connect

The navigator pops up showing all the elements of the web page that was loaded as shown in Fig. 6.32. We can make transformations like that of the worksheet that was shown Fig. 6.31.

Here, we perform the following transformations to the data loaded from the web page. To rename the file right click on the file—rename—name as 'country population' and click ok. Remove all the extra unnecessary columns and keep only the country and population columns (right click on the column name—remove). Also click on the drop down against the Country column and select only those countries that are of interest as shown in Fig. 6.33.

Next, lets connect the country population table with the Cookie sales table based on the country names as shown in Fig. 6.34. To do that click on the Cookie sales table—merge tables—click on the country column—select Country population in the drop down below—for now ignore privacy policies—click ok

An additional column named country population is created as shown in Fig. 6.35. Click on the filter option against the column and choose population—ok. The population gets displayed in the column as shown below. Then click on Close and Apply.

After this, Power BI desktop displays a report area canvas where we will create all the visuals. The right pane shows the list of tables involved under the fields tab. On the left, clicking on the grid symbol displays the table as shown previously. Clicking on the icon below the grid icon displays a model with the tables along with the relationships between the tables as shown in Fig. 6.36.

To create visualizations, click on the report icon (above the grid on the left pane). To start click on the textbox icon and add a title (Cookie company performance report). The first visualization will be to show the profit by date. Click on Cookie sales table in the right pane—select profit and date. Now, Power BI by default pulls in a bar chart. If we wish to change the type of visualization, we can go to the visualizations tab on the right pane and click on our choice. Here, we choose line chart.

Next visualization will be to show the profit by geographies. Click on country under the Cookie sales table in the right pane. Power BI automatically inserts the map visualization. Select profit from the right pane.

Next visualization will be to show the units sold by population. Click on table from the visualizations in the right pane. A blank table is inserted into the canvas. Drag and drop the required fields into the table (population, country and units sold) from the right pane.

Next, we will create a bar chart to show the sales by product. Click on the product and profit fields from the right pane and choose the bar chart from the visualizations

Fig. 6.30 Data transformation

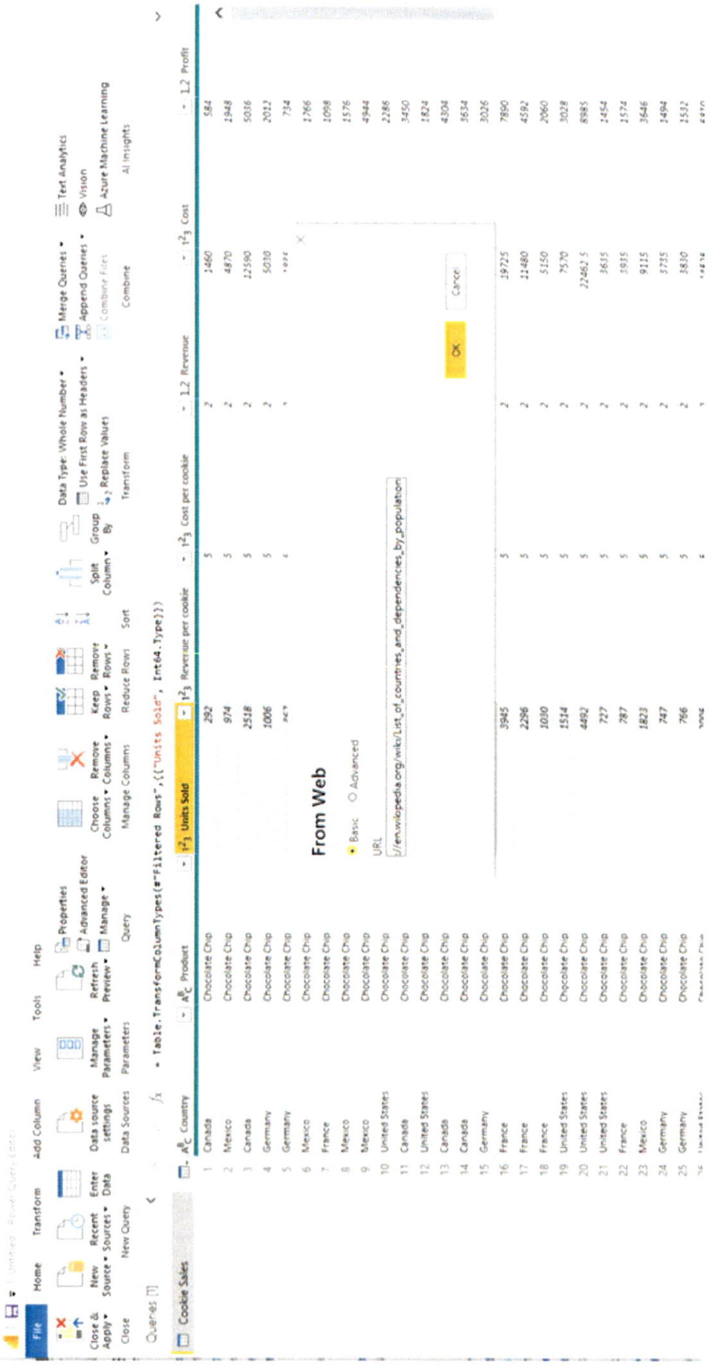

Fig. 6.31 Loading data from web

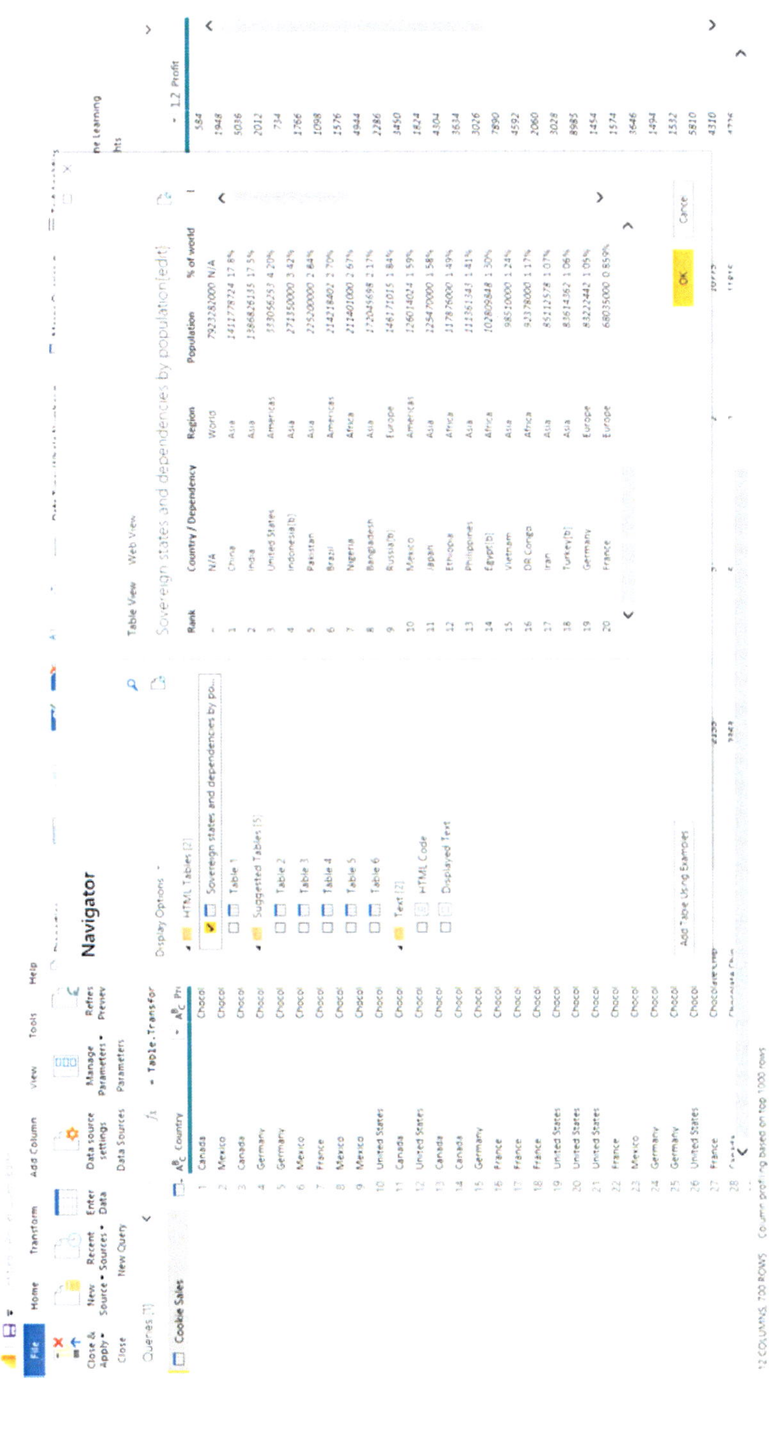

Fig. 6.32 Navigator showing elements of the web page that was loaded

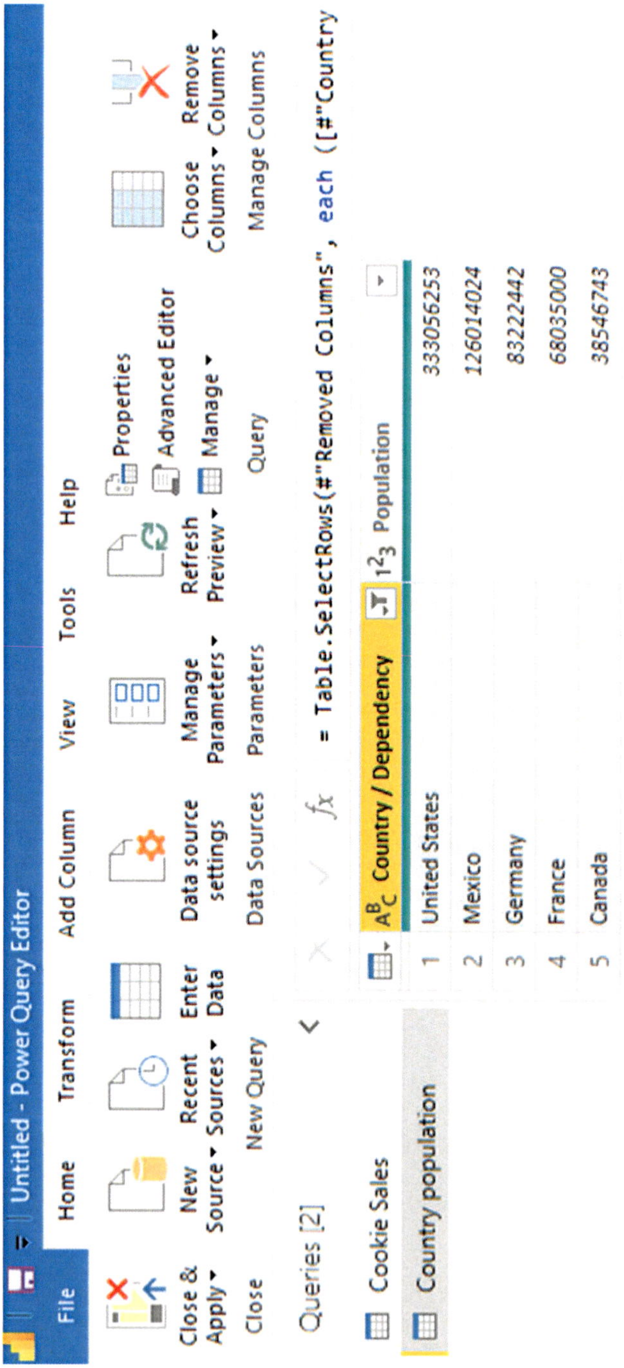

Fig. 6.33 Transformation on web data

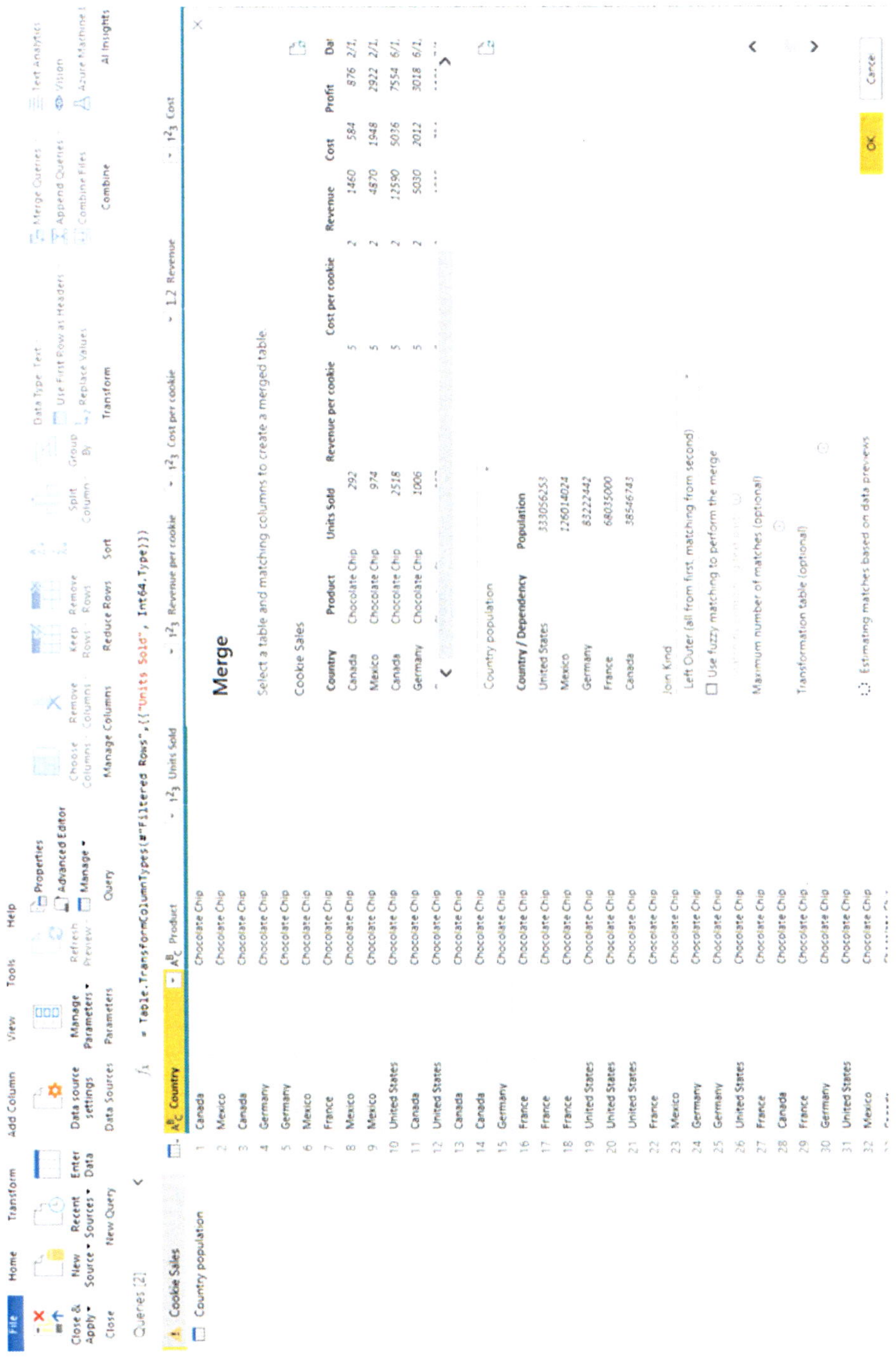

Fig. 6.34 Connecting the country population table with the cookie sales table

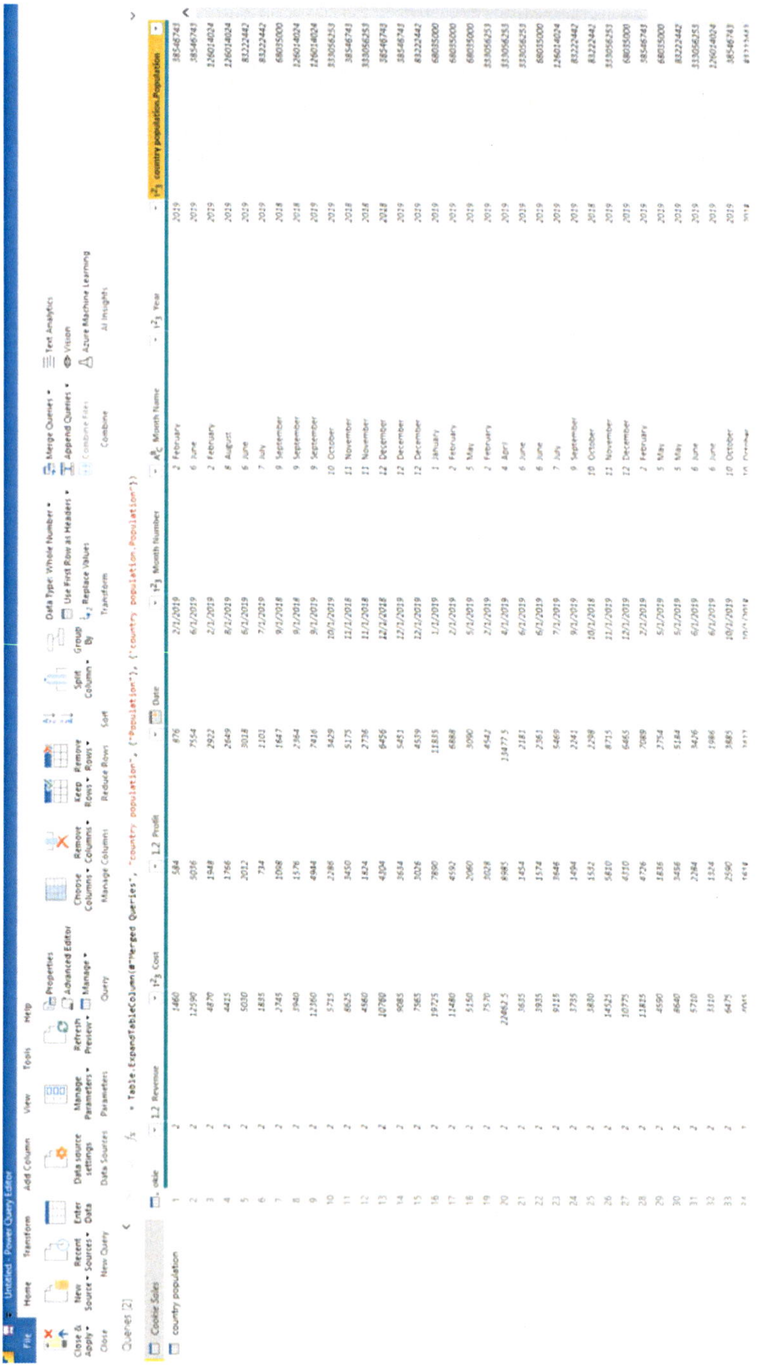

Fig. 6.35 Additional column (country population)

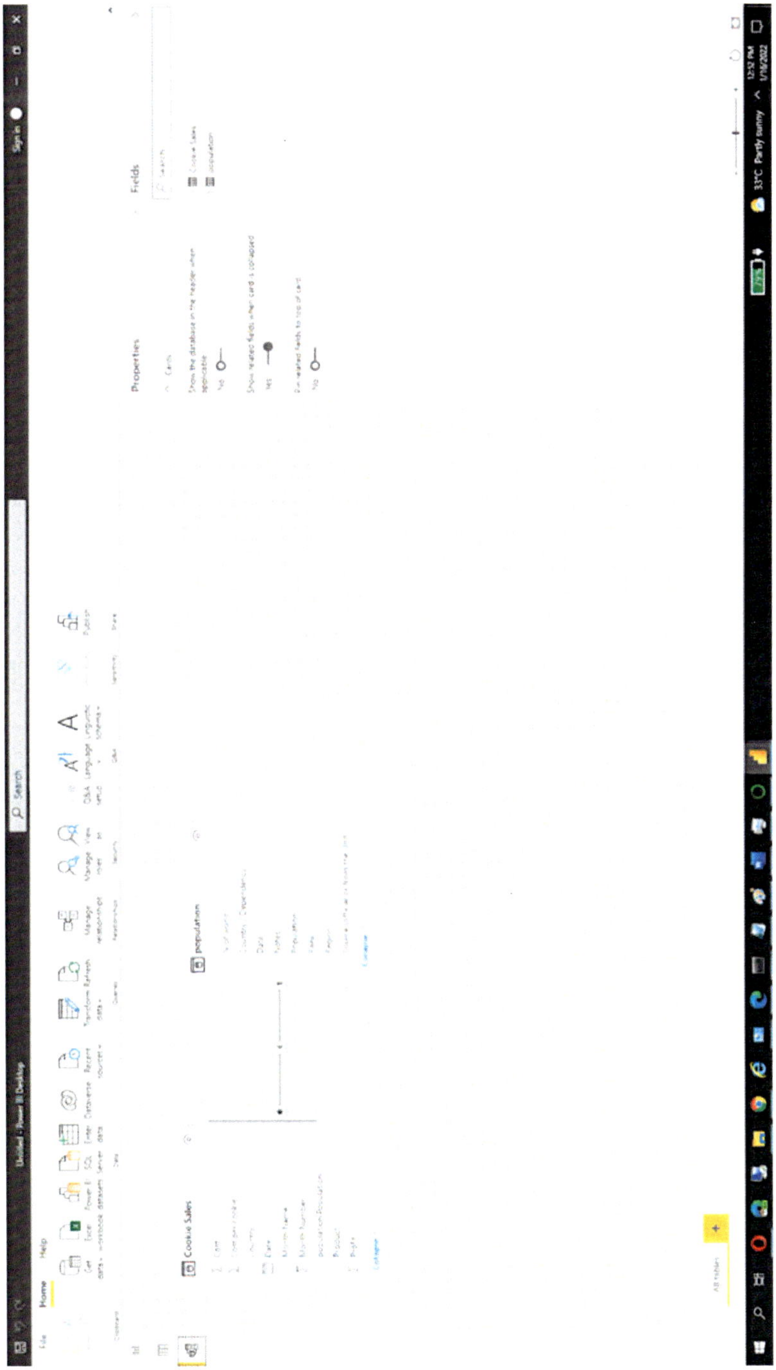

Fig. 6.36 Report area canvas

on the right pane. We can also introduce a slicer to view the charts for each country. To do this click on the slicer from the visualizations—drag and drop the country field into the slicer. Once we are done, we can save the visualizations in the required place by clicking on File—Save As—give a name and save. To publish the report and share it with others and include views from others click on File—Publish—Publish to BI—Save—choose the destination (my workspace). A dialog box indicating success on the publish activity pops up. Click on the link 'Open' 'powerBI_demo.pbix' in PowerBI to view the report in PowerBI. The visualizations' report page is shown in the Fig. 6.37.

6.2 Motivation

Data science involves the process of revealing hidden information from data collected to provide useful information using common scientific methods that help businesses better predict and make improved business decisions based on analysis and patterns. Predicting pattern and modeling functions in data science requires a good understanding of basic mathematics, as well as knowledge of using appropriate technologies such as SQL, Python, R, Matlab, Seaborn, Tableau and practical coding skills to select the right machine-learning algorithm to solve the problem. Data science has transformed the education sector where teacher and textbook dependence has plummeted. It brings together teachers and students from different places onto a common platform/ground where information sharing and job opportunities are increasing. The benefits of this flexibility were best seen in the context of the COVID-19 pandemic where students were able to continue studying online while staying safe in their homes. Data science has transformed various other fields such as banking, medical science, e-commerce and many more. To begin with, it is important to determine if machine learning or data science is the best way to solve a particular problem, or if another method is more appropriate. In addition, we must ensure that sufficient and appropriate data is available to address a specific problem, as data is the lifeblood of every data scientist. If relevant data is not accessible, the first step is to develop, retrieve and pipe the data to be stored for future use and accessibility. Also, to solve problems successfully, identifying the team with the right skills is important. Another milestone in solving data science problems is the effective solution of the challenge in terms of practical and efficient solutions and desirable performance metrics such as precision, recall and accuracy.

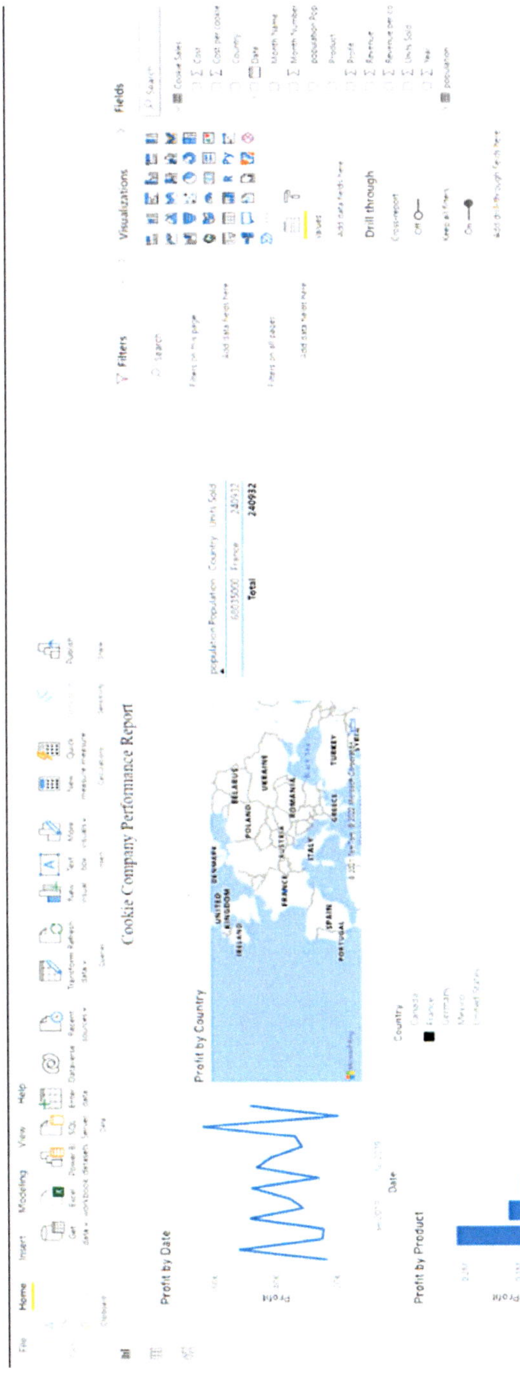

Fig. 6.37 Creating visualizations

6.3 Case Studies

Case Study 1: Biomedical Engineering Applications

The paper [9] is a biomechanical engineering application that uses data science algorithms and evaluation mechanisms to create an intelligent wearable system that uses IMUs (inertial measurement units) to identify mobility exercises and then drive the user through the exercises at home. The aim was to equip the system with capability to assess the movements of the user in minimum time and using minimum sensors while still maintaining the accuracy. To recognize this, the research goal was to obtain best possible sensor configuration and the best classifier.

The process included collection of data from a standard subject. Features we calculated for the movements and PCA (Principle component Analysis) technique were used to develop the optimal feature combination. A genetic algorithm was then applied in order to arrive at the best possible sensor configuration by iterating through the activities of signal selection, classifier training and performance evaluation of the classifiers. A classifier was developed using the features extracted and selected from the raw data. Here, supervised classifiers namely NM (nearest mean), K-NN (k-nearest neighbor), RBF (radial basis function) and SVM (support vector machine) were used and their efficiencies were compared. To evaluate the performance of the classifiers, we calculated the parameters like accuracy, specificity, precision, recall/sensitivity and F score as follows:

$$\text{Accuracy} = \frac{\text{TP} + \text{TN}}{\text{TP} + \text{TN} + \text{FP} + \text{FN}}$$

$$\text{Recall} = \frac{\text{TP}}{\text{TP} + \text{FN}}$$

$$\text{Precision} = \frac{\text{TP}}{\text{TP} + \text{FP}}$$

$$\text{Specificity} = \frac{\text{TN}}{\text{TN} + \text{FP}}$$

$F\_\text{score} = \frac{2 * \text{Recall} * \text{Precision}}{\text{Recall} + \text{Precision}}$ where, TN signifies true negatives, TP denotes true positives, FP denotes false positives, and FN represents false negatives.

In the initial step, the algorithm's performance was inspected through discovering the best configuration for only a small set of activities namely turn right, standing, seated, fast walk, turn left, slow walk, put down an object, pick up an object, stairs down and stairs up (training data set with 10 entries). On evaluating the performance of the classifiers against the parameters mentioned above, it was found that SVM and RBF were not satisfactory and hence rejected. The following table shows the outcomes of the assessment.

Classifier	NM	*k*-NN	RBF	SVM
Accuracy	1	1	0.99	0.97
Recall	1	1	0.97	0.93
Precision	1	1	0.98	0.91
Specificity	1	1	0.99	0.96
F_score	1	1	0.97	0.92
Computation duration in days	1.5	6	18	33

The k-NN and NM classifiers were selected for further rounds of evaluation. Now the algorithms efficiency was tested with 34 activities to arrive at the results in terms of best arrangement of sensors and best classifiers. Each of the algorithms were given 100 test data, and the movement recognition time was noted. The NM classifier took 2.3 s to accomplish the task, while the *k*-NN algorithm took 9.7 s. Hence, it was concluded that the NM classifier was the best fit.

With four sensors on the right arm and thigh, as well as the left forearm and shank, NM was chosen as the final classifier. A retraining algorithm was created to adjust the system to new users. Retraining the system using the new user's data, which had the same weight as the preloaded library's data, kept the classifier's precision and recall above 90%.

Case Study 2: Soil and Agricultural Engineering Applications

The project's goal [10] is to develop a model that efficiently classifies soil occurrences and maps the soil type to crop data in order to enhance prediction exactness. The model development is split up into two phases: training and testing. The soil data set and the crop data set are both used. The anticipated and actual classes are compared, and the proper classes are found in the list. Machine learning algorithms like bagged trees, weighted K-nearest neighbor (KNN), and Gaussian Kernel-based support vector machines (SVM) are used in the classification of soil series. The accuracy obtained using the KNN algorithm, Gaussian Kernel-based SVM, and bagged trees were 92.93%, 94.95% and 90.91%, respectively. Hence, it was determined that the SVM method outperformed the other two algorithms. After the classification of the soil, the crops suitable for the series are then suggested. A hybrid classifier model is used to optimize the features which is divided into three phases, namely, pre-processing, feature selection and SVM_GWO (grey wolf optimizer along with the support vector machine) to enhance the accuracy, precision, recall, and F-measure. The proposed GWO_SVM performs better than the typical SVM classification algorithm with an accuracy of 77.09%, precision of 75.38%, recall of 74.189% and F_measure of 73.15%. The model is built using the ensembling technique that combines the results from multiple machine learning algorithms such as the Naive Bayes, random forest and linear SVM to predict the appropriate crop based upon the soil type and other characteristics with high accuracy. Each of the above classifiers provide their own class labels with an agreeable frequency which is then combined using the majority voting technique. Combining the independent base learners yielded a classification accuracy of 99.91%.

6.4 Summary

The data science problem is focused on finding conclusions from available data. The real data science problem focuses on providing one of the following as outputs, i.e., separating or collecting data, pattern identification, anomaly identification, showing correlations and predicting outcomes. One of the key skills required in modeling a data science problem is the ability to accurately diagnose and interpret an existing problem.

Data science plays an important role in establishing preventive measures, predictions, recommendations and evaluation of various parameters in all aspects of life. This chapter introduces a practical approach to how problems are solved in data science through case studies. It is an add-on taken from various research papers that explain in detail the problem identification, classification and application of the mathematical model in the revised system. This will provide students with a real-life data science problem. The chapter presents an in-depth description of the required statistics for data science researchers and IT professionals. A step-by-step approach to problem-solving is presented clearly and accurately to the reader with examples for better understanding. Case studies will assist researchers and students in identifying areas where machine learning, deep learning and data science methods and tools can be helpful. A clear picture of all the machine learning strategies used for the analysis and development of models has been presented to researchers to provide a way for further development.

References

1. Trivedi KS (2008) Probability & statistics with reliability, queing, and computer science applications" PHI
2. Kothari CR (2004) Research methodology (2nd edn), New age international. ISBN(13): 978-81-224-1522-3
3. Han J, Kamber M (2001) Data mining: concepts and techniques. Morgan Kaufmann Publishers, San Francisco
4. Aho, Hopcraft, Ullman (1974) The design and analysis of computer algorithms. Addison Wesley
5. Mahalle PN, Shinde GR, Pise PJ, Deshmukh JY (2021) Foundations of data science for engineering problem solving (1st edn). Springer Verlag, Singapore. ISBN: 9789811651595
6. Mahalle PN, Nancy AP, Shinde GR, Vinayak Deshpande A (2021) The convergence of internet of things and cloud for smart computing (1st edn). CRC Press. https://doi.org/10.1201/9781003189091
7. Stančin I, Jović A (2019) An overview and comparison of free Python libraries for data mining and big data analysis. In: 2019 42nd International convention on information and communication technology, electronics and microelectronics (MIPRO), pp 977–982. https://doi.org/10.23919/MIPRO.2019.8757088
8. Sousa R, Miranda R, Moreira A, Alves C, Lori N, Machado J (2021) Software tools for conducting real-time information processing and visualization in industry: an up-to-date review. Appl Sci 11:4800. https://doi.org/10.3390/app11114800

9. Nasrabadi AM, Eslaminia AR, Enayati AMS, Alibiglou L, Behzadipour S (2019) Optimal sensor configuration for activity recognition during whole-body exercises. In: 2019 7th international conference on robotics and mechatronics (ICRoM), pp 512–518. https://doi.org/10.1109/ICRoM48714.2019.9071849
10. Rahman SAZ, Chandra Mitra K, Mohidul Islam SM (2018) Soil classification using machine learning methods and crop suggestion based on soil series. In: 2018 21st international conference of computer and information technology (ICCIT), pp 1–4. https://doi.org/10.1109/ICCITECHN.2018.8631943

Chapter 7
Pandemic Problems

7.1 Introduction

A pandemic is an outbreak of infectious disease in large scale at international level. It generally affects the large population worldwide. Any disease killing many people is not considered as pandemic. For example, worldwide many deaths are caused by cancer but is not regarded as pandemic because the cancer is not easily transmissible. For a disease to be pandemic, it must also be infectious. If a disease affects fixed number of individuals universally, then it is not considered as pandemic [1]. Reappearances of regular flu due to weather change are not considered as pandemic. In the past centuries, several outbreaks of pandemic diseases have been recorded. It includes plague, smallpox, Spanish flu, severe acute respiratory syndrome (SARS), swine flu, H7N9, Ebola, Zika and most recently COVID-19. The key features of any pandemic condition are as follows: i) effect in wide geographic region, ii) fast disease movement, iii) high attack rate and explosiveness, iv) contagiousness, v) minimal population immunity due to novelty and vi) severity [2].

The study of various pandemic problems can be done using above features. The similarities and differences can be identified among these features for various pandemic problems. The pandemic has huge negative impact on the economy, health and security of the society. It also has political and social disturbance worldwide. Figure 7.1 depicts the epidemic and pandemic overview.

As mentioned in the Fig. 7.1, **sporadic** refers to the irregular occurrence of the disease. The outbreak is at irregular time intervals, and it is scattered and isolated. When there is constant presence of disease or communicable agent in a population area, then this condition is known as **endemic.** When disease covers a specific geographic area, then it is termed as **epidemic**. **Pandemic** refers to the persistent occurrence of disease over a larger geographical area.

Increase in international movement due to tourism, suburbanization, changes in native land use and exploitation of natural resources are major reasons for the increase of pandemics. This will be continued in future with more intensity. Nations are focusing on identifying various ways to predict and limit the emerging epidemics.

P. N. Mahalle et al., *Foundations of Mathematical Modelling for Engineering Problem Solving*, Studies in Autonomic, Data-driven and Industrial Computing,
https://doi.org/10.1007/978-981-19-8828-8_7

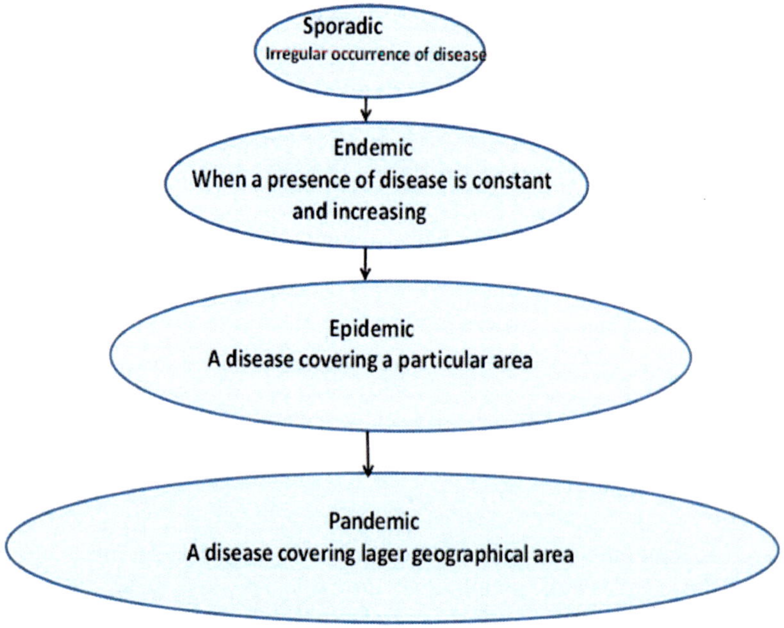

Fig. 7.1 Epidemic and Pandemic Overview

Huge investment is being done to build health infrastructure and to increase the preparedness in facing pandemic challenges [3]. As per the report released by US department of Health and Human in 2005, many countries have started devising plans to face challenges posed by pandemic. There was concern about the threat imposed by growing cases of SARS and avian, swine and other zoonotic influenza viruses. Due to the delay in reporting of SARS cases, World Health Assembly has updated the International Health Regulations (IHR) and World Health Organization (WHO) also make mandatory to all of its member countries to follow specific guidelines and standard operating procedures(SOP) for detecting, and responding to outbreaks (WHO 2005) [4]. As result of these updated IHR and guidelines given by WHO, more coordinated and effective response was given to the influenza pandemic occurred in 2009. Still there are significant lacunas in the preparedness to face the pandemic challenges at international levels. During Ebola epidemic at West Africa in 2014 gaps in the preparedness of health sector for availability of basic care, tracing of contacts at global level is exposed. There was global failure in detection of disease in time, coordination and response mobilization. Probabilistic modeling can be used to bridge these gaps in predicting the frequency and severity of pandemics. Probabilistic modeling uses big data sets of hypotheticals, scientifically plausible, simulated pandemics which includes wide range of possible scenarios. Modeling also represent the historical milestones related to developments in health care, demographical changes and changes in global travel patterns.

7.1.1 *Impacts of Pandemics*

Pandemics impacts the society in multiple ways. It has significant increase in sickness and deaths in a large geographic area and has significant impact on economic, social and political condition of the country. In this chapter, we are discussing the following issues related to pandemic problems.

- Risks
  - Throughout the human history pandemics have occurred and in recent years its frequency is increasing because of, particularly because of, the growing occurrence of virus-related disease from animals.
  - Spark risk (*where* a pandemic is likely to arise) and spread risk (*how likely* it is to diffuse broadly through human populations) are the two things which drive the pandemic risks.
  - Some parts of world (e.g., Central and West Africa) which has high spark risk are not fully prepared well to face the pandemic challenges.
  - Exceedance probability (EP) curves and other analytical tools are used for assessing pandemic risk and estimating the potential burden of pandemics.
  - Most probably, severe pandemic is caused by the influenza pathogen. In any given year, the EP analysis reports show that there is a 1% probability that influenza pandemic causes nearly $> \ = 6$ million influenza deaths globally.
- Impacts
  - Due to pandemics, there is significant increase in the mortality and morbidity at global level. It severely affects the low- and medium-income countries (LMICs).
  - Pandemics also cause the economic damages in multiple ways, which includes negative impact on the economic growth of the country.
  - Change in the individual behavior such as many persons want to remain at home and want to work from home due to fear of infection. So, there will be a significant reduction in public function and public gathering. Due to lockdown, the markets are closed. As a result of above, the economy grows negatively during pandemics.
  - Some major social and economic disruption can occur due to pandemic measures.

7.2 Motivation

The mathematical modeling of pandemic condition is used as a tool to study the methods by which viruses spread. The mathematical models are also used to foresee the future possibility of an outbreak and to check approaches for the control of an epidemic.

Any system can be expressed using model. It is bit easier to work with models instead of system. Following are the reasons for the use of models:

(a) To understand the system completely,
(b) To visualize the system's behavior and
(c) To determine the future behavior of the system.

Modeling is the one step toward system control. Various issues in modeling are as follows:

- Multiple models exist for one system
- Model is a one part of actual system. Every model has limited scope
- Modeling can be done in several ways using a variety of perception.

To predict and control the illness in human and animal modeling, simulation is used as the major decision tool. These are flexible and adjustable. However, in real situations these models need to be adjusted as per the biological characteristics of each disease, as every virus/disease has its own individual features.

John Graunt was the initial researcher who scientifically attempted to measure causes of death in his book '*Natural and Political Observations made upon the Bills of Mortality*', in 1662. The analysis done by John Graunt about the reasons of death is believed to be the beginning of the '*theory of competing risks*', and according to Daley and Gani, it is '*a theory that is now well established among modern epidemiologists*'.[5, 6]

The first mathematical model for the increase of contamination was created by Daniel Bernoulli in 1760. Bernoulli designed a model to encourage the practice of vaccinating against smallpox disease. Bernoullis' model has proved that world-wide vaccination against smallpox would increase the life expectancy from 26 years 7 months to 29 years 9 months. Daniel Bernoulli's work preceded the modern understanding of germ theory [7].

William Hamer and Ronald Ross used the law of mass action for the explanation of epidemic behavior [8].

The compartmental models are introduced in the year 1920. In 1927, Kermack et. al. has presented a model in which they considered a stable population with only three compartments. The compartmental model presented by Kermack and McKendrick (1927) and the epidemic model introduced by Reed–Frost (1928) both discus the connection between *susceptible, infected* and *recovered* or *immune* persons in a population. The Kermack–McKendrick compartmental model has successfully predicted the behavior of outbreaks which is observed in many recorded epidemics [9].

Lately, instead of compartmental models [10], the agent-based models (ABMs) becomes popular. For example, epidemiological ABMs have been used to inform public health interventions against the spread of SARS-CoV-2. Agent-based epidemic models are useful in the decisions related to mitigation and suppression measures.

7.2.1 *Types of Epidemic Models*

The pandemic models are classified as either deterministic or stochastic. The word 'stochastic' implies that there is randomness. A stochastic model is used to estimate probability distributions of potential outcomes where there is random variation in one or more inputs over time. Stochastic methods are used to estimate the statistical agent-level disease dissemination in small or large populations.

Deterministic mathematical models are used when population's size is large. Deterministic model divides individuals in population into subgroups, and each subgroup indicates a stage of the epidemic. Derivatives are used to denote the transition rate from one subgroup to another. Differential equations are used in the formulations of the model. The epidemic process is deterministic as the population size in compartment varies with respect to time Fig. 7.4.

Pandemic mathematical models are divided into 3 broad categories, Statistical methods used in pandemic surveillance, mechanistic state-space model, and empirical learning models (refer Fig. 7.2). The performance of mechanistic state-space models is better than the other two as far as description of respiratory diseases such as MERS and SARS isconcerned. Mechanistic state-Space models are further divided as 'continuum" models, 'Markov Chain' models, 'complex network' models and 'agent-based simulations' [11]. Figure 7.3 represents the states in a SIR model where β represents the rate of infection and γ represents the recovery over a period of time. S, I, and R stand for Susceptible, Infectious and Recovered respectively.

The main purpose of using the pandemic models is to monitor, to predict the outbreaks and to identify the patterns in disease characteristics so that spread of disease can be controlled [12]. Mathematical simulations can be used to assess the situation for proper resources allocation. Mathematical models and simulations are proved to be a game changer. The COVID-19 pandemic has demonstrated importance of dynamic mathematical models in prediction of epidemics.

- **Exponential Growth**

When the quantity of contaminated persons doubles in every generation, it is observed as exponential growth of epidemics, for example, if 1 person transmits disease to 2 and those 2 transmits to 4 and so on. This model is represented by the curve which grows exponentially until it reaches to zero as everyone in the population is infected by the disease. There is no peak, and herd immunity observed it is gradual decline as seen in Fig. 7.4.

- **Basic Reproductive Rate**

Basic reproductive ratio (BRR) also called as basic reproductive number is represented by $\boldsymbol{R_0}$. BRR indicates the expected number of cases directly generated by one case in a population where all individuals are susceptible to infection [13]. BRR determines whether the disease will spread sub-exponentially, vanish or remain persistent: if $R_0 > 1$, then each individual on average infects more than one other person, so the infection will spread; if $R_0 < 1$, then each individual infects fewer than one individual

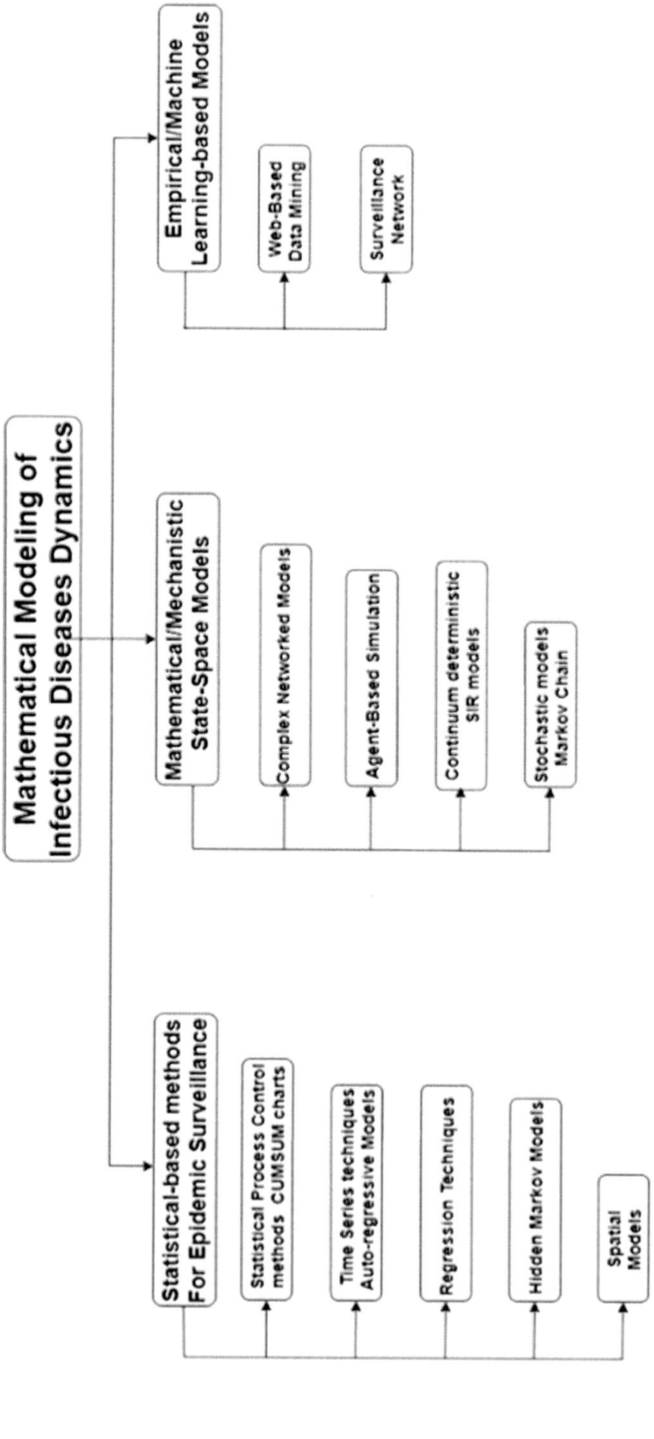

Fig. 7.2 Types of pandemic models

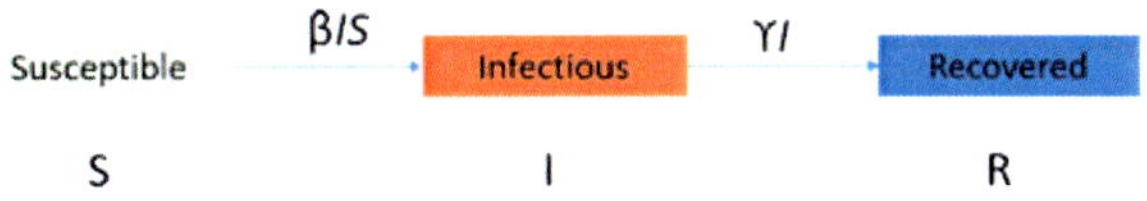

Fig. 7.3 States in SIR Model

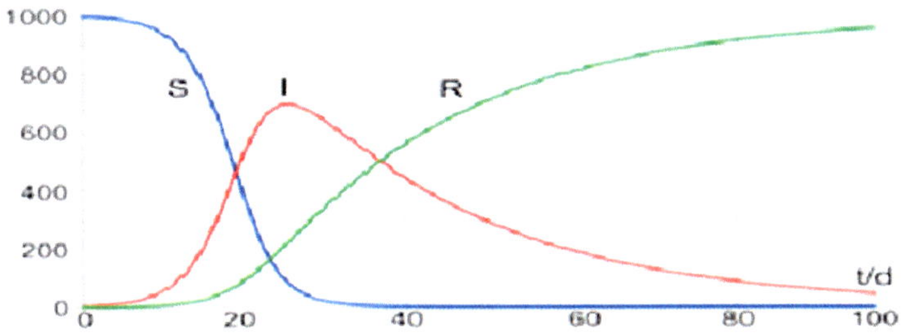

Fig. 7.4 SIR model with $S(0) = 997$, $I(0) = 3$, $R(0) = 0$ and rates for infection $\beta = 0.4$ and for recovery $\gamma = 0.04$

on average, so the disease will die out; and if $R_0 = 1$, then each individual will infect on average exactly one other individual, so the disease will become endemic: it will move throughout the population but not rise or decline.

- **Endemic Steady State**

When contagious illness is present in the population without any external inputs then it is considered as an endemic. This means that, on average, each infected individual is infecting exactly one other individual (anymore and the number of people infected will grow exponentially and there will be an epidemic, any less and the disease will die out). Mathematically it is presented as:

$R_0 S = 1.$

Product of basic reproductive ratio (R_0) of the infection, and proportion of the population that is susceptible (S) must be one with the assumption that every individual is susceptible, Note that this means that for a disease to be in the endemic steady state, the higher the BRR (R_0), the lower the proportion of the population susceptible must be and vice versa. Assume that the rectangular stationary age distribution and let the ages of infected individuals have the same distribution for each birth year. Let A is the average age of infected individuals, for instance when individuals younger than A are susceptible and those older than A are immune (or infectious). The proportion of the population which is susceptible (S) is represented by:

$$S = \frac{A}{L}$$

In the above equation, L is the age at which in this model every person is assumed to die. If we rearrange equation for the endemic steady state, then we get:

$$S = \frac{1}{R_0}$$

Therefore, as per the transitive property:

$$\frac{1}{R_0} = \frac{A}{L} \Rightarrow \mathrm{R}_0 \frac{L}{A}$$

Thus, we can conclude that basic reproduction number of a disease given A and L in either type of population distribution is given in the above equation.

7.3 Case Studies

7.3.1 Compartmental Models

When we discuss about the mathematical modeling used for infectious diseases, the compartmental models are most used modeling technique. The given population is divided into 3 categories having labels such as susceptible (***S***), infectious (***I***) or recovered (***R***). Individuals can move between the compartments. The flow patterns between the compartments represent the order of the labels, e.g., **SEIS** means susceptible, exposed, infectious, and then susceptible again.

These mathematical models for the study of infectious diseases are originated in the early twentieth century. The major contributions are from Ross[1] in 1916, Ross and Hudson in 1917,[2, 3] Kermack and McKendrick in 1927 [4] and Kendall in 1956 [14].

Most often, these models are deterministic and run well with ordinary differential equations. Sometimes it can also be used with a stochastic framework. It is related to reality and complicated to analyze.

These models are used to forecast disease spreads, total number of infections and duration of pandemic. Some other parameters such as reproduction number are also predicted using these models. These models also evaluate the effect of public health interventions on the outcome of the pandemic.

7.3.1.1 The SIR Model

The SIR model is a good example of the compartment model. In this model, population is divided into 3 categories as follows.

S: It represents the number of susceptible persons. When a susceptible and an infectious person come into contact, the susceptible individual becomes infected with the disease and then those individuals are transmitted to the infectious compartment.

I: These are the number of persons who are infectious. These are the persons who have been infected and are capable of infecting susceptible individuals.

R: is the number of individuals to be removed. We put the persons in this compartment who are either recovered or died, with the assumption that the number of deaths are negligible with respect to the total population. This compartment may also be called **'recovered' or 'resistant'.**

The predictions done by SIR model are reasonably accurate in the context of human-to-human transmitted diseases [9], and where recovery gives lasting resistance like measles, mumps and rubella (Fig. 3).

At given instance of time, the S, I and R denote the number of individuals in each category. The number of susceptible, infectious and removed persons vary over time even if the total population size remains constant. It is represented by a function of time t such as S(t), I(t) and R(t). For a particular illness in a particular population, these functions are used to estimate possible outbreaks and devise the control mechanism for it.[9]

The model is dynamic as the number of individuals in each compartment fluctuate over time which is inferred by the variable function of t. The importance of this dynamic aspect is most evident in an endemic disease with a short infectious period, such as measles in the UK prior to the introduction of a vaccine in 1968. As there is variation in the quantity of susceptible (S(t)) over time, diseases occur in the cycles of epidemics. At the time of epidemic, the quantity of susceptible cases decreases swiftly as more susceptible individuals get infected and move to infectious and removed compartments. The disease cannot reappear again until the number of susceptible has built back up.

- **Variations of basic SIR model**

SIR model discussed above is used in various pandemic problems with some variations. Following are the variations of SIR model.

The SIS model: - *Susceptible* -Infectious - Susceptible
The SIRD model: - *Susceptible-Infectious-Recovered-Deceased*
The SIRV model: - *Susceptible-Infectious-Recovered-Vaccinated*
The MSIR model: - *Maternally Derived Immunity-Susceptible-Infectious-Recovered*
The SEIR model:- *Susceptible-Exposed-Infectious-Recovered-Deceased*

7.3.2 COVID-19 Pandemic Problem

COVID-19 is also known as coronavirus disease 2019 which is an infectious disease emerged in China in December 2019. It is quickly spread globally. On 11th February 2020, the World Health Organization (WHO) gave a new name to the pandemic disease caused by 2019-nCoV as strain severe acute respiratory syndrome coronavirus 2.

Dynamics and evolution of COVID-19 are studied by various mathematical models. Three phenomenal models are presented in [12], which were validated with epidemics of other infections which are different from COVID-19. Some other works done in the [15] propose SEIR-type models with little variations, and some of them incorporate stochastic components. COVID-19 is caused by a new variant of virus.

It has created worldwide emergency situation, and we need some models which consider its specific features.

Specifically mathematical model needs to be developed which includes following.

- Model should be able to evaluate the impact of undetected infected people [13], and it should be able to predict the effect of COVID-19 on the % of detected cases over the sum of the actual infected cases. Here, θ represents the fraction of detected cases over sum of the actual infected cases. In the literature, we found various authors propose estimations of θ in several countries for the COVID-19 pandemic.
- The impact of sanitary hygienic and infectiousness conditions of hospitalized people
- To estimate the number of beds needed in the hospitals for specific regions which is the main challenge policy makers are facing during COVID-19 outbreak.

There exist various models which can be used in biological systems, but our main focus is on mathematical models [16], as they are well suited for the estimation and control of biological systems. When any specific model needs to be chosen among the available models, the general principle followed is the model frugality, which means that 'a model should be as simple as possible and as complex as necessary!'. The parsimony of the model is also an important factor for predicting the unknown factors using actual data. A more accurate model with a smaller number of parameters is preferred over a less accurate and more complex model. Measures such as the Akaike information criterion (AIC), the Bayesian information criterion (BIC) and the minimum description length (MDL) can be used to balance between the number of observations and the model unknown parameters to select between competing models with variable number of parameters and different levels of accuracy [17, 18]. Ultimately, the model is said to be useful if the parameters are interpreted properly and the model can predict the features which match the actual data.

- **Gaussian Models**

  Gaussian model (GM) is a statistical model which exhibits better performance than other models in consistently forecasting COVID-19 pandemics. It is based on the Central Limit Theorem of Statistics [14], which consists of multiple independent and random variables, and its properly normalized sum tends toward a Gaussian distribution even if the original variables themselves aren't distributed normally. In the GM, the bell-shaped Gaussian function is fitted to existing data of cumulative deaths or detected infections and extrapolate the numbers to future times [19]. The Gaussian model appears to be very simple and primitive, and its applicability can be substantially justified. Sometimes the GM can be considered as a special case of continuum models mentioned in [11]. Agent-based epidemiological model is compatible with GM [6]. In Italy to forecast the pandemic of COVID-19 virus, hybrid model was developed using Gaussian error function and Monte Carlo simulation [7]. Based upon distribution observed in the study of seasonal influenza [8], the cumulative numbers of detected cases and deaths in China were fitted with

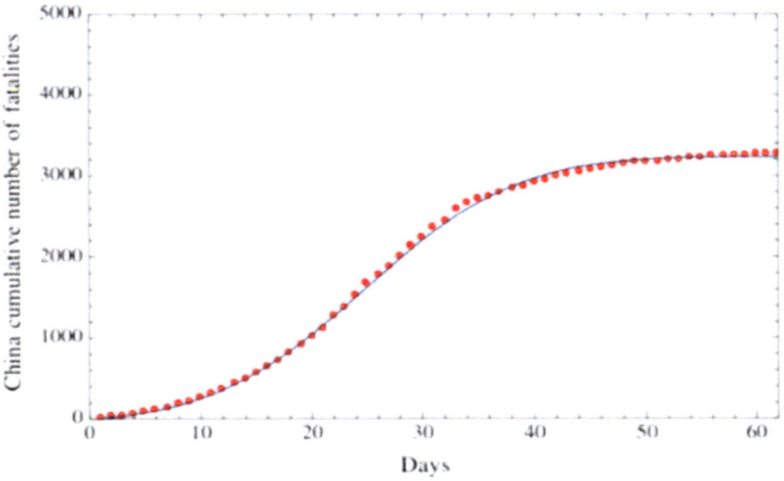

Fig. 7.5 Gaussian Error Function fitted to cumulative number of deaths caused by COVID-19 in China (red dots) from January 22, 2020 [7]

a Gaussian error function containing four parameters, refer Fig. 7.5. The Gaussian function used has been represented by following equations where erf represents Gaussian error function.

$$a + b \text{ er } f(cx + d)$$

$$\text{er } f \frac{2}{\sqrt{\pi}} \int_0^z e^{-t^2 dt}$$

7.4 Summary

As per the mathematicians, a good mathematical model reveals the data it uses. Nowadays, the main challenge faced by mathematical models is the deficiency of data in time and it does not contain any noise. The data gets corrupted mainly due to the changes in guidelines of registering the cases and variations in the symptoms of corona virus. The SIR family of models seems to be rapidly growing with ever-increasing considerations and variables, subsequently increasing the complexity of differential equations involved. Challenges faced by mathematical modelers are summarized into:

(a) **Pathogen evolution:** Most of the flu viruses responsible for the pandemics undergo mutations, which in turn disturb the model design and change the model parameters dynamically.

(b) **Statistical uncertainties**: Some not known factors such as the overall immunity of the population in the specific region affects the severity of the pandemics. There are temporal and spatial variations due to such factors. Statistical methods that extract features that describe the properties and transmission characteristics of the infection from incomplete and noisy surveillance data are yet to be developed.

(c) **Contact patterns**: Law of mass action assumes that contagious contacts between persons are purely random in nature. However, if the interaction between individuals is not random, it would result in localized saturation of infection and destabilization of these laws. Such mixing would cause complex epidemics that cannot be predicted using traditional modeling techniques.

"One model fit to all" is not applicable to devise a mathematical model in epidemiology. Proper model is identified based on the nature of the problem. The model which gives accurate result is not the proper model but the model which fulfills all the requirements of the problem addressed. The models discussed in this chapter fulfills different requirements. Some models estimate the number of deaths accurately, whereas some are good at describing the disease attributes while others are designed to explore and measure the effects of measures of intervention.

References

1. Daley DJ, Gani J (2005) Epidemic modeling: an introduction. Cambridge University Press, New York
2. Hethcote HW (2000) The mathematics of infectious diseases. SIAM Rev 42(4):599–653. Bibcode:2000 SIAMR. 42, 599H. https://doi.org/10.1137/S0036144500371907
3. Blower S, Bernoulli D (2004) An attempt at a new analysis of the mortality caused by smallpox and of the advantages of inoculation to prevent it. Rev Med Virol 14(5):275–288. https://doi.org/10.1002/rmv.443.PMID15334536.S2CID8169180
4. Hamer W (1928) epidemiology old and new. London: Kegan Paul. Ross R (1910) The prevention of Malaria. Dutton
5. Eisinger D, Thulke HH (2008) Spatial pattern formation facilitates eradication of infectious diseases. J Appl Ecol 45(2):415–423. https://doi.org/10.1111/j.1365-2664.2007
6. Adam D (2020) Special report: The simulations driving the world's response to COVID-19. Nature 580(7803):316–318. https://doi.org/10.1038/d41586-020-01003.6
7. Squazzoni F, Polhill JG, Edmonds B, Ahrweiler P, Antosz P, Scholz G et al (2020) Computational Models That Matter During a Global Pandemic Outbreak: A Call to Action. J Artif Soc Soc Simul 23(2):10. https://doi.org/10.18564/jasss.4298.ISSN1460-7425.S2CID216426533
8. Sridhar D, Majumder MS (2020) Modelling the pandemic. BMJ 369:m1567. https://doi.org/10.1136/bmj.m1567.PMID32317328.S2CID216074714
9. Maziarz M, Zach M (2020) Agent-based modelling for SARS-CoV-2 epidemic prediction and intervention assessment: a methodological appraisal. J Eval Clin Pract 26(5):1352–1360. https://doi.org/10.1111/jep.13459.PMC7461315.PMID32820573
10. Huppert A, Katriel G (2013) Mathematical modelling and prediction in infectious disease epidemiology. Clin Microbiol Infect 19(11):999–1005. https://doi.org/10.1111/1469-0691.12308.PMID24266045
11. Costris-Vas C, Schwartz EJ, Smith RJ (2020) Predicting COVID-19 using past pandemics as a guide: how reliable were mathematical models then, and how reliable will they be now? Math Biosci Eng 17(6):7502–7518. https://doi.org/10.3934/mbe

12. Nakamura GM, Cardoso GC, Martinez, AS (2020) Improved susceptible–infectious–susceptible epidemic equations based on uncertainties and autocorrelation functions. Royal Soc Open Sci 7(2):191504. Bibcode:2020RSOS....791504N. https://doi.org/10.1098/rsos.191504. ISSN 2054-5703. PMC 7062106. PMID 32257317
13. Maier BF, Brockmann D (2020) Effective containment explains subexponential growth in recent confirmed COVID-19 cases in China. Science 368(6492):742–746. Bibcode:2020Sci...368..742M. https://doi.org/10.1126/science.abb4557. PMC 7164388. PMID 32269067
14. Brauer F, Castillo-Chávez C (2001) Mathematical models in population biology and epidemiology. Springer, New York
15. Dietz K (1967) Epidemics and rumours: a survey. J Royal Statis Soc Ser A (General) 130(4):505–528. JSTOR 2982521. https://doi.org/10.2307/2982521
16. Shalizi C (2018) Data over space and time; lecture 21: compartment models (PDF). Carnegie Mellon University. Retrieved 19 Sept 2020
17. Kermack WO, McKendrick AG (1991) Contributions to the mathematical theory of epidemics–I. 1927. Bull Math Biol 53(1–2):33–55. https://doi.org/10.1007/BF02464423.JSTOR94815. PMID2059741
18. Brauer F (2017) Mathematical epidemiology: past, present, and future. Infect Dis Model 113127. https://doi.org/10.1016/j.idm.2017.02.001
19. Renz A, Widerspick L, Dräger A (2020). "FBA reveals guanylate kinase as a potential target for antiviral therapies against SARS-CoV 2" *Bioinformatics. 36 (Supplement_2):813821.* doi:https://doi.org/10.1093/bioinformatics/btaa813.
20. Nakamura GM, Monteiro, ACP, Cardoso GC, Martinez, AS (2017) Efficient method for comprehensive computation of agent-level epidemic dissemination in networks. Sci Rep 7(1):40885. arXiv:1606.07825. Bibcode:2017NatSR...740885N. https://doi.org/10.1038/srep40885. ISSN 2045-2322. PMC 5247741. PMID 28106086
21. Basic reproduction number—an overview|ScienceDirect Topics
22. Britton T, Ball F, Trapman P (2020) A mathematical model reveals the influence of population heterogeneity on herd immunity to SARS-CoV-2. Science 369(6505):846849
23. Pollard AJ, Bijker EM (2021) A guide to vaccinology: from basic principles to new developments. Nat Rev Immunol 21(2):83–100. https://doi.org/10.1038/s41577-020-00479-7
24. Vytla V, Ramakuri SK, Peddi A, Srinivas KK, Ragav NN (2021) Mathematical models for predicting covid-19 pandemic: a review. J Phys Conf Ser 1797:012009

Chapter 8
Interdisciplinary Engineering Problems

8.1 Introduction

The word interdisciplinary is coined and defined in the literature with different ways, context and meaning. The meaning of interdisciplinary is referred to multidisciplinary [1]. However, in the practical world, the meaning of interdisciplinary refers to the integration of multiple disciplines to identify the valid problem and to get the solution to same problem with the help of technology. Many times one discipline is also supported and facilitated by many sub-disciplines, and this integration is the appropriate proportion which is also called as interdisciplinary approach. Commerce in mathematics, mathematical modeling and simulation, mathematical modeling and research, mathematics for physical science, mathematics for actuarial science, etc. are some emerging areas where this interdisciplinary approach is getting popular nowadays. The interdisciplinary approach also helps to make the subject more attractive and meaningful. The key benefits of interdisciplinary problems and applying mathematics to these problems are listed below:

1. Idea synthesis and the synthesis of characteristics from many academic and research disciplines
2. Application of various academic disciplines to solve real-world problems
3. Creativity in the solution with inculcation of interdisciplinary thought process
4. Improvement in analytical and creative skills for new problems and solution
5. More opportunities and scope in the higher studies
6. Varied perceptive and critical thinking in the problem-solving
7. Product development-centric knowledge and more research rigor

P. N. Mahalle et al., *Foundations of Mathematical Modelling for Engineering Problem Solving*, Studies in Autonomic, Data-driven and Industrial Computing,
https://doi.org/10.1007/978-981-19-8828-8_8

Every use case in the modern world is driven by the tools and language from mathematics. Today, the mathematics has been transformed from the classroom discipline to the effective tool. Application areas like sciences, medicine, engineering, information communication and technologies, mathematics and modeling has become vital ingredient. Interdisciplinary problem-solving and research is getting popular day by day, and mathematics and tools is supporting many advances. Practical use of mathematics in interdisciplinary problem-solving and its mapping to the other disciplines is very important. Business and industrial applications heavily use mathematics as a language of science. Business modeling, analytics and useful inferences can be drawn if and only if the modeling of the underlined problem is done clearly. As mentioned in [2], mathematics is the meta-science and meta-arts and in the sequel referred as pivot of all sciences. Even the biology and life sciences has the great amount of base as mathematics. The impact of life sciences on quality of work-life balance, quality life in health, job satisfaction and its mathematical modeling are the modern problems in focus today. Bio-mathematics is another area which in convergence with bio-physics and bio-chemistry addresses the problems like estimation, investigations, forecasting etc. Economics and finance due to its widespread applications is driven by mathematics in the recent years. Language of mathematics which includes linear algebra, differentia equations, relations, functions and calculus is widely used in the field of economics for problem-solving.

8.2 Motivation

Interdisciplinary research, problem-solving and the related difficulties have been presented and discussed in the literature with different contexts [3, 4]. Many studies are also carried out to draw clear motivation for interdisciplinary approaches. Interdisciplinary research particularly in the doctoral programs is also addressed in [5] mentioning different experiences during initiating and facilitating it. There are many studies which are carried out to predict or to draw some meaningful inferences with the help of artificial intelligence techniques. In this case, data acquisition is one of the major challenges. Sometime the required data is generated using experimentation or it is also generated by initiating the survey. In the latter case, the data generated thorough survey is referred as categorical data [6] and the reliability of this data is the main concern. Probabilistic and stochastic theory is also used to model many problems in science and engineering which consist of uncertainty in defining the problem or there is also some amount of uncertainty in the solution. There are many studies and use cases where the stochastic theory is used in the recent problems like modeling pandemic for the purpose of forecasting, estimation, etc. [7]. Some important and interesting case studies of such interdisciplinary mathematical modeling are presented and discussed in the section below.

8.3 Case Studies

Applications of mathematics in modeling of the interdisciplinary problems are listed below:

(a) **Anniversary Decision Formulation**

Decision theory [8] provides the framework for objective of selection. Probability theory is the foundation of decision theory where the subject is not unified one. A statistical methodology, called the Bayesian decision theory (BDT) [9], uses probability to mold the collected information and arrive at decisions. BDT is a fundamental statistical approach that quantifies the trade-off between various decisions using probabilities and cost that accompany such decisions. BDT is used for decision-making under uncertainty which is simple and best suited for many uncertain problems. There is a famous and interesting anniversary decision problem [8]. You are moving back to home and suddenly recall that your anniversary is sometime in this period. It is uncertain and quite probable that it is today. You can go with roses or empty hand and accordingly there are various possible outcomes which are depicted in Fig. 8.1.

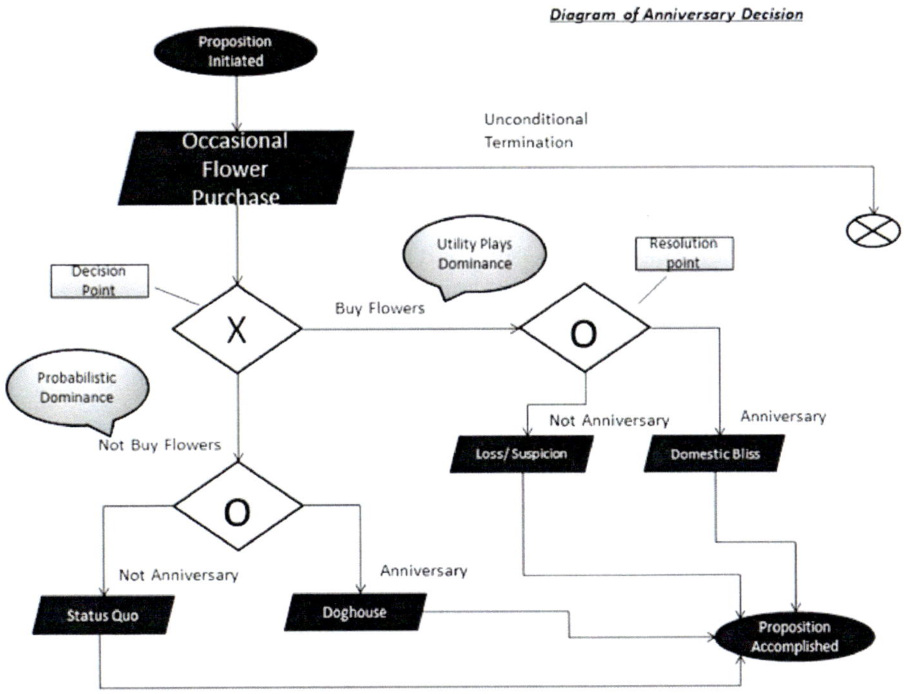

Fig. 8.1 Anniversary decision formulation

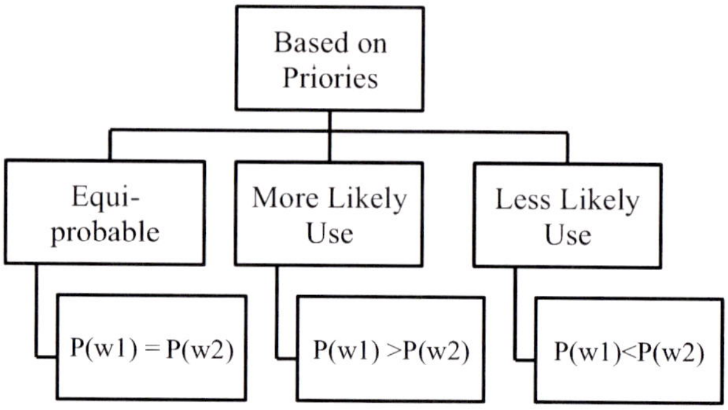

Fig. 8.2 Classification of probabilities

To take an analogy, in Internet of Things (IoT), objects in the environment fall under two set that are mutually exclusive, namely, the set for expedient objects and a set of non-expedient objects. There are different scenarios in IoT depending on three main use cases which includes indoor, outdoor and hybrid use case. Communication between one object and another occurs under two situations where the expedient and non-expedient objects have unknown probabilities, or their probabilities are completely unknown. This classification of the probabilities is depicted in the Fig. 8.2.

Based on the assumption, it is understood that the existence of other types of objects can be ignored. By the property of exclusivity, the above assumption can be represented as in Eq. (8.1).

$$P(w1) + P(w2) = 1 \tag{8.1}$$

Decisions for objects are done only if there is known beforehand as given [8] in Eq. (8.2).

$$\text{Decision (object)} = \begin{cases} w1 & \text{if} P(w1) > P(w2) \\ w2 & \text{otherwise} \end{cases} \tag{8.2}$$

This case study helps us to understand how the decision theory is useful in modeling any interdisciplinary problem where there is uncertainty in either the problem or solution.

(b) **Recurrence Relation**

Recurrence relation is a computation rule which is recursive in nature and is used to generate finite or infinite set or series. Recurrence relations can be solved either by guessing technique or by characteristic polynomial method. Consider the very famous problem of generating the Fibonacci series and the relation for the same is formulated as below:

$$F_n = F_{n-1} + F_{n-2}$$
$$F_n - F_{n-1} + F_{n-2} = 0$$

The above equation is of the form

$$a_0 t_n + a_1 t_{n-1} + \ldots + a_k t_{n-k} = 0$$

Such equations are called as homogenous linear recurrences with constant coefficient. This equation is to be solved to find out value of t_i on k for $i = 1$ to k, referred as initial conditions. This equation is linear because it contains the terms t_i with power 1, and it is homogenous because linear combination of t_i is equal to zero.

Consider the same problem of object classification in IoT environment. Typically, the process of detecting and distinguishing items that are useful or trash is a three-phase procedure which are the priori analysis followed by class conditional probability and finally posterior analysis. Priori analysis is a phase that includes execution of single step instructions. The instructions include conditional or arithmetic operations. Therefore, the time complexity involved in the first step execution is of unit time complexity. The succeeding phase includes class conditional probabilities computations. The number of feature elements of size '*b*', which could be representations of traffic, property over the network, etc., influence the calculation of class conditional probabilities. The final phase is dependent on the class conditional variables that have been calculated in the previous phase. Consider that, given the input of size say two i.e., for w_1 and w_2. Thus, the recurrence relation for time complexity obtained is as follows:

$$t_n \begin{cases} 1 & n = 0 \text{ or } 1 \\ 'b' & n = 2 \\ tn - 1 + b & n > 2 \end{cases}$$

Solving the Recurrence Relation:

Now we have an inhomogeneous recurrence relation. Thus, rewriting the recurrence relation as follows:

$$t_n = t_{n-1} + b$$
$$t_n - t_{n-1} = b$$

from above, we have $k = 1, a_0 = 1, a_1 = -1, b = b$ and $p(n) = 1, d = 0$; now forming its characteristic polynomial, we get as follows:

$$(x-1)(x-b)^{0+1}$$
$$(x-1)(x-b)^{1}$$

$$t_n = c1\,(1)^n + c2\,(b)^n$$

$$t_n = c1 + c2(b)^n$$

Given is $t_1 = t_0 = 1$, applying the result to above equation

$$c1 + c2\,(b)^1 = 1 \qquad \text{(i)}$$

$t_2 = b$, applying the result to above equation

$$c1 + c2\,(b)^2 = b \qquad \text{(ii)}$$

Solving equations i and ii, subtract ii from i:

$$c2 = (1-b)/\left(b-b^2\right)$$
$$c2 = (1-b)/b(1-b)$$
$$c2 = (1/b)$$

Substitute $c2$ in i:

$$c1 + (1/b)\,b = 1$$
$$c1 = 1$$

Thus, we have $c1 = 1$ and $c2 = (1/b)$, put these values in roots to get t_n

$$t_n = 1 + (1/b)\,(b)^n$$
$$= 1 + (b)^{n-1}$$
$$\boldsymbol{t_n \sim (b)^{n-1}}$$

Hence, it can be concluded the above recurrence relation has the time complexity of the order of $O((b)^{n-1})$, where b is the size for the set of feature element 'x'. This example shows that how recurrence relation is useful to formulate the mathematical model of any interdisciplinary problem. In addition to this, by solving such recurrence relation also helps to derive the time complexity of the problem mathematically.

(c) Optical Signal-to-Noise Ratio-based Clustering Strategy for Underwater Wireless Optical Sensor Networks

The story of underwater wireless networks begins when humans first found that a sound wave can travel through water. Decades later, it was experimented that even optical waves can pass through water medium. The research community has even proved that optical waves are most efficient underwater communication medium but only for smaller distances (10 to 50 kms). The term efficient is in means of bandwidth, data reliability and data security. The efficiency is due to unconditional properties of optical wave in underwater environment. For example, minimum absorption coefficient confirms data reliability. The low attenuation coefficient helps in achieving higher bandwidth but maximum scattering coefficient leads to smaller distance travel. The underwater optical communication medium is in demand as two advantageous things overcome one drawback.

The rapid developments in underwater applications for different sectors have also led to the popularity of underwater wireless optical sensor networks (UWOSN). Therefore, there is an increasing researcher's attention with a motive of emphasizing more on improvements in UWOSN. There are mainly three factors responsible for an efficient and effective network. These factors are data reliability, data security and network performance. As discussed above, we know that reliability and security are well achieved due to natural properties of optical waves. The minimum attenuation and absorption make UWOSN reliable, and quantum cryptography with qubits in optical communication medium helps in achieving data security.

The most important issue or research gap is with third factor, i.e., 'network performance', which depends on total number of active nodes, embedded software efficiency and total energy consumption. This study begins here and aims to develop a performance-aware green network, which will reduce or save the energy of optical nodes. The underwater nodes battery is difficult to recharge or replace, and so energy is precious resource in underwater wireless networks. In order to achieve energy efficiency, clustering is one of the most widely used techniques. The clustering strategies used for Terrestrial Wireless Sensor Networks [TWSN] and Underwater Wireless Sensor Networks [UWSN] are mainly based on machine learning. In state of the art, the study suggests that ML-based clustering strategies are higher in time and space complexities. In this view, we proposed a clustering strategy for UWOSN based on Optical Signal to Noise Ratio [OSNR]. The OSNR values help in selecting an appropriate cluster head.

Definition OSNR: The ratio of optical signal power to noise power, suggesting stronger signal power as compared to noise error. As per the OSNR definition, the higher the OSNR value, the healthy is an optical node. The word healthy is in terms of lesser data retransmissions and connectivity with higher data rates.

Note: The OSNR formula used in proposed clustering strategy is also self-derived.

Proposed Performance-aware Green Clustering Algorithm

OECA Algorithm

1. Inatilization of network parameters.
2. Declaration of clustnum = 0,n = 0, Ao = threshold value.
3. Energy calculations per nodes, i.e., Ai where i = 1,2,3,4……n
4. if (Ai > = Ao), then
   {
   Add that node, say m to set P {j1,j2,j3……….jm}
   }(i! = n)
   end if
5. For (I = 0; I < = n, I + +)
   {
   OSNR calculation per node in set P { }, by below formula
   $O_k = (\text{OSNR})_k = P_o e^{-c(\lambda)z}(t_k + s_k + h_k + w_k)$
   where k € P,}
   end for
6. Table creation of nodes and their OSNR values.
7. Select the first maximum O_k value form table and denote it as (OM)1.
   $(\text{OM})1 = \text{Max}\{O_k | \text{k} € \text{P}\}$.
8. Elect node b as cluster head and broadcast HELLO message.
9. Nodes sending response message are consider as cluster members.
   CL1 = {x |b-x| ≤ 30 m}.
   Suppose |CL1| = r1.
10. Delete r1 nodes of CLn from P
11. increment clustnum + +
12. if (clustnum! = optimal clusters number)
Repeat from step 3, until P = null
13. Data Sensing()

Mathematical Modell for Clustering Strategy

1. Selection of Cluster Head

   Let 1,2,3,… *n* be *n* nodes
   Let *Ai*, *i* = 1,2,3… *n* be the energy of ith nodes
   And
   Ao be the threshold value of energy
   Let *Aj*1, *Aj*2, *Aj*3,… *Ajm*, be m nodes having energy greater than or equal to *Ao*.
   Let P={*j*1, *j*2, *j*3,... *jm*}
   Therefore, find the OSNR of the nodes which are in T by following formula
   $O_k = (\text{OSNR})_k = P_o e^{-c(\lambda)z}(b_k + c_k + t_k + v_k)$ for all k € P
   Now, select the node having maximum value of
   Let $(\text{OS})1 = \text{Max}\{O_k | k € P\}$.
   Suppose (OS)1 corresponds to node b € *P*
   Therefore, node b is the cluster head of the first cluster
   CL1 = {*x*|*b*-*x*| ≤ 30 m}
   Suppose|CL1| =*r*1
   Now delete r1 nodes of CL1 from *P*.

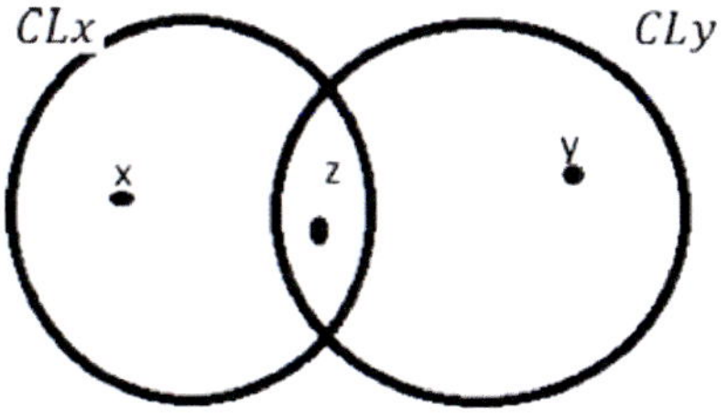

Fig. 8.3 Venn diagram of overlapping clusters CLx and CLy

$P = P - \mathrm{CL}1$
Apply same procedure for nodes in $P1$, $P2$, $P3$, …Pn
And find (OS)2,(OS)3,… (OS)n

1. Selection of Cluster Member

Continue the same procedure until we get Pp= $\emptyset$
If z be any node in CLx $\cap$ CLy $\neq \emptyset$
Then find the distance between z and cluster heads of CLx and CLy say x and y respectively as shown in Fig. 8.4
These are following three cases.
a: If |x–z|>|y–z|, then z € Cy, so remove z from $_{\mathrm{CL}x}$.
b: If |x–z|<|y–z|, then z € CLx, so remove z from CLy
c: If z is non-active node, then it is not eligible to become cluster member of any cluster.

This use case proves that how simple tools and language from mathematics can be used to model interdisciplinary problem. In order to do this, it is very important to understand the interplay between mathematical modeling and interdisciplinary problems.

8.4 Summary

Science, technology, engineering and management are attracting interdisciplinary research nowadays. Analyzing real-life use cases and situation with the help of mathematics helps to transform the entire process to enable rich learning and understanding. Mathematical modeling of interdisciplinary problems requires applied mathematics knowledge, and in the sequel, some case studies are presented and discussed in the scope of this chapter. An application of conditional probability in anniversary decision problem is presented in the first part. The second case study explains how recurrence relations are useful in the formulation of mathematical modeling of interdisciplinary situations and in turn solving these recurrences results into the time analysis. The last case study presented and discussed in this chapter explains how mathematics is useful for deriving modeling of some research use case.

References

1. Dey N, Bhateja V, Hassanien AE (2016) Medical imaging in clinical applications algorithmic and computer-based approaches. Springer
2. Sidhu KS (1995) The teaching of mathematics, (Fourth Edition). Sterling Publishers, New Delhi, India
3. Lynch J (2006) It's not easy being interdisciplinary. Int J Epidemiol 35:1119–1122
4. Frodeman R, Klein JT, Mitcham C (2010) Oxford handbook of interdisciplinarity. Oxford University Press, Oxford
5. Pfirman S, Martin P (2010) Facilitating interdisciplinary scholars. In: Frodeman R, Klein JT, Mitcham C (eds) Oxford handbook of interdisciplinarity (pp 387–403). Oxford: Oxford University Press
6. Shinde GR, Majumder S, Bhapkar HR, Mahalle PN (2022) "quality of work-life during pandemic", studies in big data, vol 100. Springer, Singapore
7. Shinde GR, Kalamkar AB, Mahalle PN, Dey N, Chaki J, Hassanien AE (2020) Forecasting models for coronavirus disease (COVID-19): a survey of the state-of-the-art. SN Comp Sci
8. North DW (1968) A Tutorial Introduction to Decision Theory. Syst Sci Cybern IEEE Trans 4(3):200–210
9. Bayes TR (1763) An essay towards solving a problem in the doctrine of chances," philosophical transactions of the royal society of London 53, 370–418 (reprinted with biographical note by G. Barnard, 1958, in Biometrika 45 293–315).
10. Parikshit N, Mahalle NR, Prasad Prasad R (2013) Object classification based context management for identity management in internet of things. Int J Comput Appl 63(12)

Chapter 9
Conclusion

9.1 Summary

This section concludes the book and proposes the future outlook which can be explored further to initiate new research avenues. This book intends to cover the need of mathematical modeling, the key motivation behind it and the different types of problems in engineering and the appropriate mathematical tools required to model them. Today, the mathematical modeling has become an integral part of the research journey and it brings more impact to the research as well as problem formulation. As we come to the conclusion on discussion of mathematical modeling, the book brings out to the reader a bag full of overview of mathematical modeling to model the problems and techniques to improve on the research and contributions.

The book specially mends for novice researchers gives them a feel and understanding of how effective mathematical modeling can be to bring more impact. The tremendous growth in the need of research in engineering discipline and some thumb rule of modeling the problem with mathematics is the key contributions in the book. During the process of mathematical modeling, it is very important to understand the type of problem and analyze it. The mathematical modeling is also explained with the different use cases in the book. Every chapter represents the type of problem in engineering, its overview, the motivation of having discussed this type of problem and the selection of appropriate to model the problem with case studies. The book makes budding researcher to consider these factors and emphasizes to be thoughtful in the process of mathematical modeling. To summarize, it is all about bringing out more impact in the problem formulation using mathematics and building the model for it!!!

P. N. Mahalle et al., *Foundations of Mathematical Modelling for Engineering Problem Solving*, Studies in Autonomic, Data-driven and Industrial Computing,
https://doi.org/10.1007/978-981-19-8828-8_9

9.2 Research Openings and Future Outlook

Due to increasing need of research in engineering and bring in more quality publication of the repute, mathematical modeling, its increasing complexity with the varied problem and particularly data intensive research is going to be a major research area for next few decades. The key engineering domains like artificial intelligence, machine learning and deep learning have become interdisciplinary areas and research in all branches of engineering is applying these domains to solve the problem. The prominent research areas which can be explored further are listed below.

- Explainable artificial intelligence for driving use cases.
- Performance validation and modeling of problems using explainable artificial intelligence.
- Predictive analytics for preventive approach.
- Data exploration at edge computing.
- Data exploration at fog computing.
- Prescriptive analytics.
- Data modeling.
- Performance evaluation of different data models.
- Mathematical modeling of Internet of things use cases.
- Industrial Internet of things.
- Automobile internet of things.
- Internet of things for agriculture.
- Internet of things in Pharmaceuticals.

Decision problems, optimization problems network problems, data science problems, and interdisciplinary engineering problems, its mathematical modeling, the selection of appropriate mathematical tools for modeling and the explanation of modeling process with different case studies are the key contributions of this book. In addition to this, the stackup analysis in mechanical engineering, under water wireless sensor communication, the Internet of behaviors in web computing, edge and fog computing, and piping techniques in chemical engineering are few important key areas which can be explored further in the context of mathematical modeling. As a takeaway from the book, one must be motivated to look into futuristic opportunities in this world of problem formulation and mathematical modeling.